Springer Series in Advanced Manufacturing

The **Springer Series in Advanced Manufacturing** includes advanced textbooks, research monographs, edited works and conference proceedings covering all major subjects in the field of advanced manufacturing.

The following is a non-exclusive list of subjects relevant to the series:

1. Manufacturing processes and operations (material processing; assembly; test and inspection; packaging and shipping).
2. Manufacturing product and process design (product design; product data management; product development; manufacturing system planning).
3. Enterprise management (product life cycle management; production planning and control; quality management).

Emphasis will be placed on novel material of topical interest (for example, books on smart manufacturing and sustainable manufacturing) as well as new treatments of more traditional areas.

As advanced manufacturing usually involves extensive use of information and communication technology (ICT), books dealing with advanced ICT tools for advanced manufacturing are also of interest to the Series.

Springer and Professor Pham welcome book ideas from authors. Potential authors who wish to submit a book proposal should contact Anthony Doyle, Executive Editor, Springer, e-mail: anthony.doyle@springer.com.

INDEXED IN SCOPUS

Muhammad Hammad · Hassan Ali ·
Ghulam Hussain · Xiaoping Wang

Industry 5.0: The Human-Centric Future of Industry

Springer

Muhammad Hammad
Faculty of Aerospace Manufacturing Engineering
Nanjing University of Aeronautics and Astronautics
Nanjing, China

Hassan Ali
Prometheus, Higher School of Technology and Management
Polytechnic Institute of Viana do Castelo (IPVC)
Viana do Castelo, Portugal

Ghulam Hussain
Department of Mechanical Engineering
University of Bahrain
Isa Town, Bahrain

Xiaoping Wang
Faculty of Aerospace Manufacturing Engineering
Nanjing University of Aeronautics and Astronautics
Nanjing, China

ISSN 1860-5168 ISSN 2196-1735 (electronic)
Springer Series in Advanced Manufacturing
ISBN 978-3-032-12621-4 ISBN 978-3-032-12622-1 (eBook)
https://doi.org/10.1007/978-3-032-12622-1

This Springer imprint is published by the registered company Springer Nature Switzerland AG
The registered company address is: Gewerbestrasse 11, 6330 Cham, Switzerland

Dedicated to my beloved mother

—Muhammad Hammad

To Muhammad Ali Khan, Shehnaz Begum, Sara, Roomi, Sawaira, Laiba, Zaid, and Shaheer

—Hassan Ali

To my Childs

—Ghulam Hussain

To my siblings

—Xiaoping Wang

Foreword

The narrative of industrial evolution has always been one of human ingenuity, technological advancement, and societal transformations. From the steam engines of the 18th century to the digital systems of the 21st, each industrial revolution has redefined the relationship between humans, machines, and the environment. Today, as we stand at the dawn of Industry 5.0, we are witnessing a profound shift, one that re-centers the human being at the heart of industrial innovation.

This book, *Industry 5.0: The Human-Centric Future of Industry*, arrives at a pivotal time when the world is searching for balance between automation and empathy, efficiency and ethics, productivity and purpose. It recognizes that technology alone cannot drive progress; rather, progress is sustained when human creativity, compassion, and critical thinking guide technological power. Through its comprehensive structure; spanning the historical evolution of industries, the enabling technologies, human-machine collaboration, sustainability, data and cybersecurity, and the real-world applications shaping tomorrow. This book serves as both a map of transformation and a manifesto for responsible innovation.

By exploring key technologies such as artificial intelligence, robotics, digital twins, and blockchain and contextualizing them within human-centric design, ethics, and sustainability, this book provides a clear vision of how organizations can embrace Industry 5.0 to create resilient, inclusive, and sustainable value systems. Its inclusion of real-world case studies from manufacturing and healthcare to agriculture and smart cities demonstrates how the principles of Industry 5.0 are already reshaping industries around the globe.

Ultimately, the message is clear: the future of industry is not just about smarter machines, but about smarter collaboration between humans and technology. Industry 5.0 calls for a future where innovation serves humanity where progress enhances not only productivity but also the quality of life and the sustainability of our planet.

This book will be an invaluable resource for students, researchers, engineers, policymakers, and business leaders seeking to understand and shape this next era of industrial transformation. It stands as a timely reminder that the most advanced industry is one that never forgets its human heart.

October 2025

Prof. Dr. Haiyan Zhao
University Professor
Department of Mechanical Engineering
Tsinghua University
Beijing, China

Preface

Being on the threshold of a new industrial revolution, we need to understand how powerful Industry 5.0 can be. The fast development of technological advances has transformed our attitude to production, business processes, and the structure of society. The next step of evolution, Industry 4.0, which is enabled by the Internet of Things (IoT), Artificial Intelligence (AI), and automation, has predetermined a future that is not only more connected but also more human-oriented. Industry 5.0 is the next logical step in this continuum, where a combination of human beings, machines, and technology merge into a more sustainable, more resilient, and more inclusive future.

The book reflects on the premises of Industry 5.0, starting with a historical background of the evolution of industrial revolutions from the dawn of mechanization (Industry 1.0) to the digital era (Industry 4.0). The knowledge about these earlier revolutions is included in Part I. It is priceless in terms of gaining insight into the forces that are causing the current shift and the lessons that have been learned in the process. Part II of the book explores deep into the core technologies of Industry 5.0. Whether it is collaborative robots (cobots) and AI-enhanced decision support, digital twins, or smart environments, the technologies covered in this book are the core of this new age, as they allow for reaching previously unheard of heights of productivity, personalization, and efficiency.

Industry 5.0 is human-centric, and this book devotes a whole Part III to discussing the importance of workforce transformation, ergonomics, and upskilling as key to the success of Industry 5.0. With the redefinition of the human role in the workplace by Industry 5.0, the importance of continuous learning and human-machine collaboration increases even more. It is not only a technological change but a human-oriented one that is interested in the well-being and empowerment of the workforce.

Another important aspect of Industry 5.0 is sustainability and the circular economy, and Part IV discusses the way in which these values are interconnected with technological progress. Industry 5.0 is not just about the enhancement of the production capacity but doing it in a sustainable way. Intelligent processes, circular economy systems, and regulatory systems will make sure that industries become sustainable and efficient in resource usage.

There is no doubt that data, cybersecurity, and privacy are essential in this new industrial environment. As the information flows through the interconnected systems, it is essential to guarantee the safety and integrity of the information. Part V discusses the problems and resolutions in data analytics and cybersecurity, as well as shows how the 5.0 industry is being implemented in various sectors, such as manufacturing, healthcare, agriculture, and smart cities.

Although the prospects of Industry 5.0 are incredible, it is not a smooth-sailing adoption. The last part of this book involves the challenges that industries encounter in the transition process and how to overcome these challenges. We present the case studies and success stories of how early adopters have overcome these challenges and provide a case that can be used as a lesson to others. The strategies to convert to industry 5.0 are also discussed in this part. In our vision of the future, we think of the consequences of the new technologies that are coming and the vision of Industry 6.0 in the long term.

The book is not merely a technical excursion but also a guide to how the wider society will change with Industry 5.0. As we embark on this exciting new world, we need to know how we can use these innovations to make the world a sustainable, inclusive, and truly human world.

As we continue through the pages, we would like to encourage the readers, leaders, innovators, and thinkers to not just welcome the opportunities that Industry 5.0 opens up to them but also be aware of the challenges that await them. Industry 5.0 is a collective endeavor, and its success will depend on how we navigate its complexities together as a global society.

Nanjing, China — Muhammad Hammad
Viana do Castelo, Portugal — Hassan Ali
Isa Town, Bahrain — Ghulam Hussain
Nanjing, China — Xiaoping Wang

Acknowledgements The first acknowledgement goes to Dr. Muhammad Hammad, whose dedication and relentless effort have made the completion of this book possible. His invaluable contributions and vision in the field of Industry 5.0 have been the driving force behind this work. Dr. Hammad not only brought this book to life but also provided the guidance and insight that shaped its direction. It was his pioneering spirit and commitment to excellence that made this journey both possible and meaningful. His previous book on Industry 4.0 was a great inspiration and that book motivated him to write the book on Industry 5.0.

The credit goes to Prof. Ghulam Hussain and Prof. Wang Xiaoping for their professional expertise and assistance throughout this process. Their insightful feedback, meticulous editing, and constant encouragement helped refine the manuscript and elevate it to its final form. Without their support, this book would not have achieved its present quality.

A special acknowledgment goes to Dr. Hammad's mentor, Prof. Dr. Wasim Ahmed Khan, whose guidance during the creation of Dr. Hammad's first book laid the foundation for this work. Our sincere appreciation to the GIK Institute for fostering an environment of academic excellence that allowed this book to come to fruition.

Sincere appreciation to Hassan Ali for his invaluable contribution to this book. His dedication and efforts were crucial in shaping some of the key sections of this work.

Competing Interests The authors have no competing interests to declare that are relevant to the content of this manuscript.

Contents

About the Authors

Dr. Muhammad Hammad is a post-doctoral research fellow in the Faculty of Aerospace Manufacturing Engineering, Nanjing University of Aeronautics and Astronautics, China. He completed a bachelor's and master's degree in the Faculty of Mechanical Engineering from the University of Engineering and Technology, Lahore. Following this education, he achieved a Ph.D. degree in the Faculty of Mechanical Engineering from the Ghulam Ishaq Khan Institute of Engineering Sciences and Technology, Topi, Swabi, Khyber Pakhtunkhwa, Pakistan. As a researcher deeply immersed in the nuances of Industry 4.0, Smart Manufacturing, and the Security of the Industrial Internet of Things (IIoT), he has carved a niche in these rapidly evolving disciplines. He has already authored two books *Smart Manufacturing Blueprint: Navigating Industry 4.0 Across Diverse Sectors* and *Computer Aided Engineering Design and Manufacturing: A Fourth Industrial Revolution Perspective.* As an engineer, researcher, and scholar, he embodies the spirit of inquiry and innovation that drives the modern world of engineering and technology.

Hassan Ali is a doctoral researcher in Mechanical Engineering at the University of Aveiro, Portugal, affiliated as a research fellow with the ProMetheus/UIDM Research Unit at the Polytechnic Institute of Viana do Castelo (IPVC). He holds a master's degree in mechanical engineering from the Ghulam Ishaq Khan Institute of Engineering Sciences and Technology (GIKI), Pakistan, obtained in 2023, and a bachelor's degree in industrial engineering from the University of Engineering and Technology, Peshawar, where he graduated with distinction and received the University Gold Medal in 2021. Hassan's research experience centers on metal additive manufacturing (MAM) particularly Laser Powder Bed Fusion (LPBF), and numerical process simulations. He has co-authored several peer-reviewed journal articles, conference papers, and textbooks in the field. Beyond research, Hassan actively contributes to academic writing, European research proposal development (including Horizon Europe initiatives), and cross-institutional collaboration in Industry 5.0, digital twins, and human-centric manufacturing systems.

Dr. Ghulam Hussain is a full professor of Mechanical Engineering. He has a rich experience of working both in academia and industry. He is a renowned scholar of manufacturing with expertise in Additive Manufacturing, Hybrid Manufacturing, Sustainable Manufacturing, Metal Forming, Welding, Material Characterization, Process Modelling and Optimization, and Surface Engineering. He is the author of over 145 scholarly articles and the editor of several books published by Taylor & Francis. He has received exceptional citations (5550+) and H-index (41) in a short span of his career. He has been recognized as a leading researcher—among the top 10 in the field of Incremental Forming, and as the top 6th National Researcher in Engineering. Since 2021, he has been consistently listed as one of the world's top 2% scientists. He is regarded as a distinguished scholar in many top tier universities and is actively engaged in joint research there. Professor Ghulam is also working as an editor and as a board member of reputed international journals.

Xiaoping Wang is a professor at the School of Mechanical and Electrical Engineering, NUAA. He earned his Ph.D. degree from the Department of Aerospace Manufacturing Engineering of Northwestern Polytechnical University in 2002. He once served as a visiting professor at Arizona State University in the United States. He is currently a consulting expert of the Chinese Academy of Engineering and a member of the editorial board of the international journal *Current Engineering Letters and Reviews*. He mainly engages in research on digital design and manufacturing, automatic fiber placement/tape laying for composite materials, robot path planning, and optimization of mathematical models related to intelligent manufacturing. He has won a total of three awards at or above the provincial and ministerial level. Three authorized and publicly released invention patents have been obtained, three books have been published, and over 100 academic papers have been published, among which 63 have been indexed by SCI and EI.

Part I
Introduction and Historical Context

Chapter 1
Introduction to Industry 5.0

1.1 Objectives

- To explain the meaning of the Industry 5.0 and how it is differentiated with Industry 4.0.
- To discuss the major principles of Industry 5.0, such as human-centricity, resilience and sustainability.
- To examine the motivation factors in the process of implementing the Industry 5.0 paying attention to the economic, social, and environmental aspects.
- To discuss the equilibrium between automation and human creativity in the ethical responsibility in Industry 5.0.
- To gain insight into the wider context and consequences of Industry 5.0 on the industries all over the world especially the aspect of innovation, sustainability and social impact.

1.2 Industry 5.0: Definition and Scope

Industry 5.0 is the next step in industrial progress, the implementation of the lessons of the Industry 4.0 and a shift in the direction of a more human-oriented, creative and sustainable industrial model. Although Industry 4.0 was more about automation, connectivity, and integration of the Internet of Things (IoT), artificial intelligence (AI) and robotics, to develop highly efficient and connected systems, Industry 5.0 proposes a cooperative model in which human and machine collaborate to boost production processes, as opposed to the replacement of human labor by machines. This transition lays more stress on human creativity, innovation and sustainability [1, 2].

Human–Machine Collaboration: The main difference between Industry 5.0 and the previous version of it, the Industry 4.0, is the focus on human–machine cooperation.

M. Hammad et al., *Industry 5.0: The Human-Centric Future of Industry*, Springer Series in Advanced Manufacturing, https://doi.org/10.1007/978-3-032-12622-1_1

Industry 4.0 is characterized by automation and robotization that was aimed at efficiency, speed, and profitability at the expense of human participation in most of the tasks. Nevertheless, Industry 5.0 appreciates that although machines are inevitable when it comes to high efficiency, they should collaborate with humans to ensure that they are more creative and have enhanced problem-solving skills [3].

AI and Robotics in Industry 5.0: Unlike the conventional robotic systems which take over jobs of human workforce in certain activities, the collaborative robots (cobots) in Industry 5.0 are developed to collaborate with human workforce to increase productivity, creativity and innovation. Cobots help humans in carrying out repetitive or dangerous duties and humans work on jobs that involve judgment, decision making, emotional intelligence and creative thinking.

Example: In production, cobots help humans to build complex parts where the human creativity and problem-solving ability is used with precision and speed of the robots where it creates a personalized production that responds to the needs of individual customers.

Focus on Human Creativity and Innovation: Industry 5.0 recognizes the fact that human ingenuity is the key to innovation and the development of solutions that cannot be achieved through technology. With the help of the synergy of human and machine interaction, Industry 5.0 aims at providing employees with tools and technologies to innovate and customize goods and services at large volumes [4–7].

Creative Collaboration: AI, automation, and other high-tech solutions will be used to simplify the work and help in analyzing and optimizing data, but human employees will be in the focus of the creative process. Industry 5.0 is centered on design thinking, personalization, and innovation, and technology, in this case, does not play the defining factor compared to the human aspect of creativity.

Example: In fashion design, AI may be used to create design alternatives depending on consumer preferences, but human designers add creativity, cultural and emotional attachment to the end product so that it is appealing to consumers and reflects public values.

Ethical Responsibility and Sustainability: One of the most significant shortcomings of Industry 4.0 was the fact that it played with efficiency, profit maximization, and automation and went too far without paying enough attention to human dignity, ethics, and sustainability. These gaps relate to Industry 5.0, which focuses on the connection of ethical responsibility and environmental sustainability to the industrial process [8–10].

Ethical AI: Industry 5.0 advocates the creation and application of ethical AI and that the decisions made by intelligent systems are within human values, societal requirements, and the legal framework. This goes to include making AI systems transparent, fair and accountable.

Sustainability: Industry 5.0 is concerned with green manufacturing and the circular economy to reduce waste and reuse resources and design items so that they can be recycled or remanufactured. It uses technology to make the best use of resources and minimize emissions and sustainable manufacturing methods.

Example: In car manufacturing, companies can employ AI-based systems to minimize the amount of energy used during the manufacturing process, limit the amount of emissions, and make sure that the materials are used are recyclable and sustainable.

1.2.1 The Range of Industry 5.0

Industry 5.0 is more than the technological innovation that was observed in Industry 4.0. Whereas Industry 4.0 was mainly focused on enhancing the efficiency of operations, connectivity and the use of data to make decisions, Industry 5.0 has gone further to consider human values as well as the effect it has on society and sustainability throughout the production chain and business processes [11].

Human dignity and ethical AI: Industry 5.0 aims at producing inclusive systems that guarantee human dignity and equal opportunities, as well as the equality of AI and automation technologies, which should not be used to increase inequality in society. The ethical frameworks are also put in place to determine the use of AI to serve humanity and not to exploit or marginalize some groups of people.

Sustainability and the Circular Economy: Industry 5.0 embraces sustainability as the core of industrial activities, the focus on efficiency of resources, recycling, and low-emission manufacturing. It brings green technologies into the scene that lower the environmental impacts and do not decrease or even enhance industrial production.

Holistic Integration: Industry 5.0 is all about the integration of high technologies, such as artificial intelligence, internet of things, and robotics, in all the aspects of the production process, including manufacturing, healthcare, agriculture, and even service industries. The systems will be more flexible, smart and sensitive to human needs and environmental challenges.

Global Change: The effects of Industry 5.0 are not limited to a sector or industry but are set to change the industries around the globe and which will produce more customized products, enhance quality of life and come up with more innovative solutions to the global challenges.

Example: In healthcare, Industry 5.0 may use AI and collaborative robotics to help doctors in surgery and guarantee accuracy with the possibility of human supervision and compassion in treating a patient.

Figure 1.1 shows a worker, wearing a safety helmet, interacting with a robotic arm in a factory environment. It represents the integration of human labor and automation, emphasizing collaboration between humans and machines.

Fig. 1.1 Human interaction with automation

1.3 The Main Principles of Industry 5.0

Industry 5.0 is a major development in the previous industrial revolutions, since it no longer focuses on automation and digitalization but on the inclusion of humans, robustness, and sustainability. The main principles of the industrial ecosystem are to establish a system that does not only concentrates on efficiency and technological development but also prioritizes human values, creativity, and environmental responsibility in the industrial activities.

i. **Human-Centricity**: The main feature of the Industry 5.0 is that it is human-centered, and a human being is not regarded as an operator or a part of the automated system but a full-fledged partner of highly developed technologies. This principle is opposed to the Industry 4.0 which was greatly concerned with automation and substitution of human work by machines. Industry 5.0 has placed human beings at the center of industrial operations and created a place where there is collaboration between machines and human beings to attain higher productivity and creativity.

   Working with Cobots: The use of collaborative robots (cobots) is one of the most prominent features of Industry 5.0 as these robots are not supposed to substitute for people, they are meant to work with them. Cobots take over all the tasks that are repetitive, unsafe, or physically demanding, whereas humans do those that involve creativity, critical thinking, and emotional intelligence.

Creativity and Problem-Solving: With this model, humans have the capability of using their problem-solving skills, judgement, and emotional intelligence to make complex decisions in real-time which cannot be done by the machines alone. This enables the workers to perform more strategic and creative roles in the production process and their value contribution goes beyond productivity.

Personalization and Consumer Demand: The cooperation of human and AI-based systems can lead to the mass customization process, and the consumer needs can be satisfied with highly personalized products. Such production versatility gives the business the opportunity to satisfy more consumers and engage them in the business as well as the ability of the workers to concentrate on areas that specifically need their skills.

Illustration: AI can forecast trends in the fashion industry and assist in generating designs according to preferences of consumers, designers can use cobots and their creativity to transform those designs into personalized collections.

Powerful Workforce: Industry 5.0 creates an atmosphere of an empowered workforce by promoting human participation in decision-making and the creative process. The transition promotes job satisfaction because employees will have the opportunity to concentrate on the job that they love and that they are good at, instead of being assigned to trivial duties.

ii. **Resilience**: Another critical value of Industry 5.0 is resilience, which appeared as the result of the experience gained by the global community in the context of the pandemic (COVID-19). Whereas Industry 4.0 was concerned with efficiency and optimization, which in many cases took the form of sacrificing flexibility and adaptability, Industry 5.0 is meant to construct resilient systems capable of responding effectively to the changes and challenges that are unpredictable.

Living with Disruptions: Industry 5.0 focuses on designing production systems that can live with disruptions, e.g. market fluctuations, supply chain shocks and global crisis. The trick to resilience is to design systems that can make changes to the operations without significant changes in production and quality.

Self-Learning Algorithms and Predictive Analytics: Resilience in Industry 5.0 is facilitated with the help of self-learning algorithms and predictive analytics. They keep track of the work and learn with the information to foresee possible interruptions, so that the industries can make changes in production or supply chain, not as a reaction, but before it is too late.

Flexible Production Systems: Industry 5.0 encourages the development of flexible and modular production systems, which can be reconfigured without any difficulty in addressing the changing requirements. Such flexibility will make businesses remain operational even when there are sudden changes in the market or external forces [12].

Example: During the pandemic, the sector, which implemented more resilient systems, including the supply chain management systems driven by AI and flexible automation systems, could shift the production lines or manage inventories better, as they could ensure continuity despite significant disruptions [13].

iii. **Sustainability**: Industry 5.0 is based on sustainability, and it takes into consideration the increasing need of the world to be more environmentally friendly in production and development. Although Industry 4.0 introduced enormous advantages regarding productivity, it also pointed out the necessity to minimize the environmental impact and utilize the resources more efficiently by industries.

Green Technologies: Industry 5.0 promotes the use of green technologies in industrial manufacturing systems and attempts to decrease the carbon footprint of the industrial manufacturing process. The usage of AI and IoT systems allows optimizing the use of resources and tracking energy consumption to reduce waste and emissions.

Circular Economy: Industry 5.0 is founded on the concept of circular economy by which products are intended to be reused, remanufactured, and recycled. Instead of adhering to the conventional linear production process (take, make, dispose), Industry 5.0 aims at developing systems in which resources are recycled and re-utilized throughout the whole product lifecycle.

Reduction of Waste and Optimization of Resources: The use of robots and AI systems is implemented to optimize energy consumption, handle waste, and separate the material to be recycled. Industry 5.0 promotes the utilization of renewable sources of energy, such as wind and solar power, in the manufacturing sector, whereby the industrial development will not be at the cost of the environment.

Eco-Friendly Production: Industry 5.0 is also concerned with eco-friendly production by emphasizing the use of low impact materials, minimizing emissions and closed loop systems so as to decrease environmental degradation and improve or maintain the same efficiency in production [14].

Example: The automotive industry the AI-powered systems may be used by the manufacturers in the automotive industry to maximize the use of materials and energy in the production process to make sure that the waste is as minimal as possible, and the end products are recyclable.

Figure 1.2 illustrates the interconnectedness of human workers, automation technologies, and sustainability in Industry 5.0. It emphasizes the synergy between these elements to drive innovation and environmental responsibility in the industry.

1.4 Factors Behind the Industry 5.0 Adoption

The implementation of Industry 5.0 is driven by several important factors, each of them reacting to the increasing pressure of the need to develop a more human, robust, and sustainable model of industry. The focus of these drivers is on economic pressures, social expectations and environmental concerns. With the dynamics of

Fig. 1.2 Key principles of Industry 5.0

the world changing ever so rapidly, industries are gradually leaning towards the principles of Industry 5.0 in order to counter these issues.

i. **Economic Pressures**: The economy of the world is changing, and the companies are under more pressure to remain competitive, enhance their profitability and adjust to the changing market needs. Since automation and technology are advancing, firms are struggling to achieve the right balance between investment in technologies and the employment of human resources. The solution to the issue is to find a way of combining high technologies and human creativity in order to increase productivity without the loss of the value of the workforce, which is also what Industry 5.0 is all about.

   Cost Reduction and Efficiency: There is pressure on companies to curtail the cost and to enhance efficiency of the operations. Although automation and AI have helped lower the cost of production in such industries as manufacturing, the human factor could not be neglected when it comes to creativity, individuality, and customer interaction. The paradigm of Industry 5.0 is not as focused on pure optimization of speed and profit as it is more even-handed and appreciates the importance of human contribution to innovation and strategic decision-making.

   Example: In terms of consumer electronics, automation may be utilized to increase efficiency and cut down costs in terms of assembly lines, but human workers are important in terms of design and customization of such products guaranteeing that they match customer preference and changing market trends.

   Labor Shortages and Skill Gaps: Labor shortages and skill gaps have been plaguing the industries around the world, but with the help of Industry 5.0, which enhances human abilities, it can become a solution. Through robotics, AI, and automation systems, companies can increase the capacity of their workforce to perform more complex tasks thereby making the workforce more flexible and capable so as to respond to emerging challenges.

Example: In agriculture, a lack of labor in the countryside can be reduced by using autonomous farming machines and AI-powered crop surveillance systems where fewer workers can operate bigger and more complicated farms and concentrate on more skillful work like farm management and innovation.

ii. **Social Expectations**: The consumers today are highly aware of the social and environmental implication of the purchases they make. With the world becoming more aware of such issues as social equity, fair labor practices, and environmental sustainability, more and more people require businesses to employ ethical manufacturing processes. The answer to these social demands is Industry 5.0 that incorporates the notions of corporate social responsibility (CSR) and ethical responsibility at all levels of production [15].

Ethical Manufacturing and Corporate Responsibility: Industry 5.0 focuses on incorporating ethical values into the manufacturing process, making corporations responsible to the society, and adhering to the standards of fairness, inclusiveness, and sustainability. Businesses are expected to do more than just produce quality products and make profits to ensure that they do good to society and practice ethics.

Examples: Patagonia is one of the brands that are adding the principles of ethical sourcing, fair wages, and sustainable production to their business models in the fashion industry. Industry 5.0 contributes to such efforts with the help of tools based on AI and robotics, which enable streamlining production processes and maintaining their ethical responsibility.

Emergence of Conscious Consumers: As a result of the increasing popularity of conscious consumption, companies are now under pressure to conduct their operations with the social good as the driving force behind their values. Industry 5.0 has adopted this in a way that has encouraged transparency and sustainability such that businesses serve the ethical demands of the contemporary customer.

Example: Food processing firms in Industry 5.0 put emphasis on the sourcing of their food and fair trade, as customers demand transparency regarding how their food is produced as well as its origin.

iii. **Environmental Concerns**: The necessity to fight climate change and overcome the problem of the degradation of the environment has resulted in the world tendency to increase eco-friendliness of the practices in all the spheres, including the industry. The answer to this demand is Industry 5.0 that is aimed at environmental sustainability and encourages the application of green technologies, energy-efficient production processes, and the principles of the circular economy to minimize the effects of industrial processes on the planet [16].

Minimization of Carbon Footprints: Industry 5.0 is promoting the importance of minimizing carbon footprint through adoption of renewable energy, minimization of energy usage and use of environmentally friendly technology. It emphasizes production procedures that will limit waste, emissions and enhance resource utilization during the product lifecycle.

Example: In the automobile industry, the makers of electric vehicles (EV) are considering cutting carbon emission by using sustainable materials and energy-efficient production. AI and robotics are useful in streamlining the production process with the least amount of energy consumption and waste without compromising on the quality of products.

Circular Economy and Resource Conservation: Industry 5.0 will encourage the transition to a circular economy that will reuse products, remanufacture them, and recycle them. This is opposed to the traditional linear model in which the resources are taken out, utilized and discarded. Industry 5.0 will combine closed-loop systems, which will allow companies to reuse the materials, limit waste, and make sure that the production process leaves the minimum environmental footprint.

Case study: Industry 5.0 can help in the production of electronics products that are easy to recycle and remanufacture to minimize e-waste and environmental pollution of discarded electronic products.

Green Technologies: Industry 5.0 also encourages the use of bio-degradable products, renewable sources of energy and recyclable items in the industry. The monitoring and optimization of production can be achieved using AI and robotics that will make sure that the industries decrease their ecological footprint and continue to be productive.

Example: The industries in construction and building materials are using sustainable materials more and more including recycled concrete and biodegradable insulation. Such technologies as AI-based supply chain management and cobots can facilitate such processes and make them efficient and sustainable.

Figure 1.3 outlines the key drivers for the adoption of Industry 5.0, including economic pressures, social expectations, and environmental concerns, highlighting the growing need for sustainability and innovation in industry 5.0 practices.

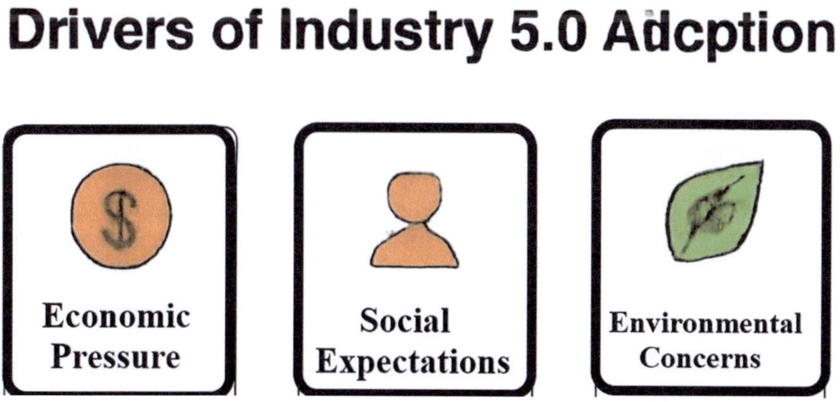

Fig. 1.3 Drivers of Industry 5.0 adoption

1.5 Need to Balance Between Automation and Human Creativity and Ethical Responsibility

Among the most important features of Industry 5.0 is that it tries to find a balance between the growing need to be more automated and the unsurpassable value of human creativity. Although automation and artificial intelligence (AI) have transformed the industries, Industry 5.0 is not aimed at replacing human beings but to incorporate them into the automated systems where they can make use of the machines. This middle ground can be utilized so that the ingenuity of human beings is utilized, and the efficiency and precision of automation is enjoyed.

Industry 5.0 is a change in the way of thinking. It is the acknowledgment of the power of automation and the importance of human input. Under this method, humans and machines supplement each other and this increases productivity, innovativeness and ethics in the industrial world [17].

- **The collaboration rather than replacement**: Among the greatest distinctions between Industry 5.0 and Industry 4.0, there is the change of replacement to collaboration. Robots and AI in Industry 4.0 were mostly regarded as an instrument to substitute for the human workforce, especially in routine or hazardous jobs. Although this resulted in improved efficiency, it also brought the issue of displacement of jobs and human factor in production. Industry 5.0, in its turn, has a significant focus on the collaborative environment, with robots and AI being viewed as partners that complement human abilities instead of substituting them. This transformation is being fueled by the fact that human labor has creativity, decision making skills and emotional intelligence in the workplace which is key to innovations and problem-solving.

  Cobots (Collaborative Robots): Introduction of cobots is an important part of such collaborative model. Cobots are to be used in collaboration with human beings, and they can handle the physically demanding, repetitive, and even dangerous tasks. Cobots will enable humans to concentrate on activities that have higher value, including critical thinking, strategy, and creativity by relieving them of these chores. This results in a more motivated workforce and a satisfying work ethic whereby human beings are no longer left to perform repetitive duties.

  Example: In the automotive industry, cobots can help people in the assembly process, including moving heavy components or solving a repetitive welding process, whereas humans can conduct quality assurance, design, and troubleshooting. Such cooperation leads to more effective production and better products.

  Job satisfaction and productivity: The model not only enhances productivity but also enhances job satisfaction since workers are able to focus on the tasks that need higher order of thinking and can be of more value to the success of the organization [18].
- **Creativity and Innovation**: Although AI and automation allow handling some tasks more effectively, they remain unable to demonstrate the same creativity and critical thinking as humans do. Industries that require innovation, which include

the designing of products, strategic decision-making, and problem-solving, need human creativity. This is particularly in the industries that require customization, flexibility and innovativeness to succeed [19–22].

Industry 5.0 focuses on the fact that human beings cannot be substituted by machines in the process of creative thinking and innovation. Instead of regarding automation as the substitute of human jobs, Industry 5.0 promotes symbiosis between human imagination and robot efficiency. Through this relationship, it is possible to have personal products, enhanced problem solving and the development of ideas over time.

Human Contribution to Innovation: Humans play an important role in the sectors where out-of-the-box thinking, strategic thinking and emotional intelligence are important. Although AI and automation may help to deal with routine work, human workers are the ones with vision and creative power to solve complex problems, create new products, and optimize the current processes.

Example: In fashion industry, AI can be used to predict trends and come up with a first draft of the designs using consumer data, but human designers bring in their creativity, cultural sensitivity and artistic touch to the final product. This partnership leads to the creation of new collections that attract the customers in a more profound way.

Customization of Products: The industry 5.0 also allows production of highly customized products using the combination of both humans and machines. As an example, AI can be used to determine consumer preferences and make forecasts, whereas human designers are able to develop solutions that suit the needs of an individual [23].

- **Ethical Responsibility**: As more people embrace AI and automation, there is a rising concern to ensure that the ethical issue of these technologies is addressed. Industry 5.0 is highly focused on ethical responsibility, which means that technological progress is not going against the values and well-being of society. It is important to make sure that such systems work transparently, accountably, and fairly as AI is increasingly used in decision-making.

  Ethical AI: The significant issue of AI is that it may cause bias in algorithms and that the decision-making process is not transparent. Industry 5.0 provides a solution to this by making sure that AI systems become ethical, accountable, and in line with the values of society. It encompasses the removal of biasness in algorithms, making the decisions made by AI transparent and explainable, and the fairness and inclusivity in all the decisions made as far as technology is concerned.

  Data Privacy and security: As the use of data grows in AI and automation systems, there is a worry of data privacy and security. Industry 5.0 makes sure that companies use personal data in a responsible way and they follow the rules regarding privacy and do not violate the rights of humans. Being ethically responsible also implies that one should not use data in a way that interferes with privacy and exposes people to danger.

  Job Displacement and Workforce: One of the most important ethical issues of automation is the possible displacement of employees. Industry 5.0 is an answer to this challenge, as it places a lot of emphasis on reskilling, upskilling, and retraining programs to ensure that workers are ready to work in jobs that exploit the best of both worlds, technology and humans. Industry 5.0 is also aimed at the development of new employment in AI maintenance, management of robotics, and AI ethics control.

As an example, with robots and AI systems replacing the routine work, human workers can shift to tasks that need more human control and judgment. As an example, in healthcare, AI may play the role of performing diagnostics and analyzing data, and doctors and nurses will be involved in the care of the patients, and technology is not supposed to substitute the human factor [24–28].

Inclusive Practices in Tech Adoption: Industry 5.0 guarantees that the process of adopting new technologies is inclusive, and anyone regardless of their backgrounds and levels of skills can have a chance to be involved in the process [29].

Figure 1.4 emphasizes the importance of balancing automation with human elements like creativity and ethical responsibility. It highlights collaboration over replacement, the role of creativity and innovation, and the need for ethical responsibility in the process.

The concept of Industry 5.0 aims at finding the compromise between automation and human creativity to make sure that technological improvements are intended to help improve, but not to substitute the human workforce. With its focus on collaboration instead of replacement, creativity and innovation, and ethical responsibility, Industry 5.0 is opening the path of a more human, innovative, and responsible future in the industrial practice [30].

The human–machine collaborative model in Industry 5.0 not only increases productivity but also makes sure that the human being will always be in the center of decision-making, problem-solving, and innovation. Such a balanced method, based

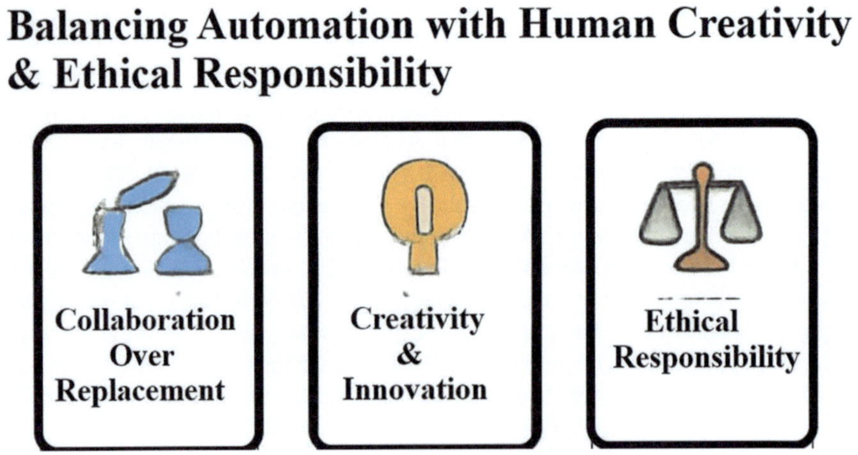

Fig. 1.4 Balancing automation with human creativity and ethical responsibility

on ethical AI, employment, and inclusive policies will lead to the development of a workforce that will perform well in a technologically advanced and socially responsible setting.

1.6 Learning Outcomes

At the end of this chapter, the readers will be in a position to:

- Learn about the Industry 5.0 and the way it is developed based on the past industrial revolutions.
- Knowing the main principles and impetuses leads to the implementation of Industry 5.0.
- Examine the role of human creativity, robot collaboration and automation in Industry 5.0 and the ways they can be used to improve productivity and innovation.
- Explain the ethical aspects of Industry 5.0 such as its devotion to sustainability and adoption of responsible technology.
- Assess how the Industry 5.0 will affect the different sectors, such as manufacturing, healthcare, and business operations, with the emphasis on human well-being and environmental conservation.

1.7 Review Questions

1. What does the change in Industry 4.0 to Industry 5.0 mean to the relationship between humans and machines at work?
2. What are the ethical issues of the greater use of AI and automation of Industry 5.0? What are the ways of reducing these challenges with the help of ethical AI and human-centered design?
3. How can Industry 5.0 contribute to the flexibility of industrial systems to the global shocks like an economic crisis or pandemic? Give practical examples to back your answer.
4. Examine how human dignity and well-being fit in the industrial processes of Industry 5.0. What makes this human-focused approach distinguishable between the previous generations of Industry and the engagement of workers and their satisfaction with the job?
5. What is the role of green technologies and principles of the circular economy to minimize the environmental impact of industrial production within this paradigm?
6. Which are the major economic and social factors that lead to the usage of Industry 5.0?
7. As Industry 5.0 is becoming more and more focused on personalization, how do AI-powered systems and collaborative robots (cobots) allow the manufacturing of personalized goods?

8. Can the automation of systems and self-learning algorithms in the Industry 5.0 be used to assist companies in coping with unexpected disruptions in the supply chain?

Appendix: Solution Manual

1. **What does the change in Industry 4.0 to Industry 5.0 mean to the relationship between humans and machines at work?**
   Solution: The relationship between humans and machines in Industry 5.0 is collaborative rather than replacement-driven. Machines perform repetitive tasks, while humans focus on complex decision-making, problem-solving, and innovation. For example, in a car manufacturing facility, cobots assist with assembly, while human workers oversee quality control and customization, ensuring that the final product meets customer-specific needs.
2. **What are the ethical issues of the greater use of AI and automation of Industry 5.0? What are the ways of reducing these challenges with the help of ethical AI and human-centered design?**
   Solution: Ethical challenges include algorithmic bias, data privacy concerns, and job displacement. Mitigation strategies involve implementing transparent algorithms, ensuring data privacy policies are enforced, and retraining workers for new roles. Ethical AI frameworks should guide the design of systems that are inclusive and equitable.
3. **How can Industry 5.0 contribute to the flexibility of industrial systems to the global shocks like an economic crisis or pandemic? Give practical examples to back your answer**
   Solution: Industry 5.0 enhances resilience by incorporating flexible production systems, predictive analytics, and AI-driven decision-making. For example, during the COVID-19 pandemic, some industries used AI to optimize production and supply chain management in response to sudden demand fluctuations, ensuring continuity in operations.
4. **Examine how human dignity and well-being fit in the industrial processes of Industry 5.0. What makes this human-focused approach distinguishable between the previous generations of Industry and the engagement of workers and their satisfaction with the job?**
   Solution: Human dignity and well-being are central to Industry 5.0, where humans engage in creative and strategic tasks while automation handles repetitive jobs. This improves job satisfaction and worker empowerment. For instance, employees in Industry 5.0 environments are involved in design and innovation, resulting in greater job fulfillment and less monotonous work.
5. **What is the role of green technologies and principles of the circular economy to minimize the environmental impact of industrial production within this paradigm?**

Solution: Industry 5.0 integrates green technologies such as renewable energy sources and recyclable materials into industrial practices. The circular economy model ensures that products are designed for reuse and recycling, minimizing waste. For example, companies in the automotive industry are incorporating electric vehicle technologies, which reduce carbon emissions.

6. **Which are the major economic and social factors that lead to the usage of Industry 5.0?**

   Solution: Economic drivers include the need for efficiency, cost reduction, and enhanced competitiveness, while social drivers focus on ethical responsibility and consumer preferences for sustainable products. These drivers shape corporate strategies toward investing in automation, AI, and green technologies, and meeting consumer demand for eco-friendly and ethically produced goods.

7. **As Industry 5.0 is becoming more and more focused on personalization, how do AI-powered systems and collaborative robots (cobots) allow the manufacturing of personalized goods?**

   Solution: AI-driven systems enable mass personalization by analyzing customer data to produce customized products, while cobots allow for flexible, small-batch production without sacrificing efficiency. In the fashion industry, for instance, AI can help design clothes based on individual preferences, while robots assist in the assembly of custom garments.

8. **Can the automation of systems and self-learning algorithms in the Industry 5.0 be used to assist companies in coping with unexpected disruptions in the supply chain?**

   Solution: Autonomous systems and self-learning algorithms allow companies to predict and adapt to disruptions in supply chains. For example, during supply shortages, AI can automatically adjust inventory levels and production schedules, ensuring that critical products are available while minimizing waste.

Bibliography

1. Maddikunta PKR et al (2022) Industry 5.0: a survey on enabling technologies and potential applications. J Indus Inform Integr 26:100257
2. Khan WA et al (2025) Industry 5.0. In: Computer aided engineering design and manufacturing: a fourth industrial revolution perspective. Springer, Cham, pp 523–535
3. Ghobakhloo M et al (2023) Behind the definition of Industry 5.0: a systematic review of technologies, principles, components, and values. J Indus Prod Eng 40(6):432–447
4. Khan WA et al (2025) Computer aided manufacturing. In: Computer aided engineering design and manufacturing: a fourth industrial revolution perspective. Springer, Cham, pp 283–297
5. Khan WA et al (2025) Production planning and control. In: Computer aided engineering design and manufacturing: a fourth industrial revolution perspective. Springer, Cham, pp 425–453
6. Khan WA et al (2025) Manufacturing and production. In: Computer aided engineering design and manufacturing: a fourth industrial revolution perspective. Springer, Cham, pp 65–185
7. Khan WA et al (2025) Pillars of Industry 4.0 and CAD/CAM. In: computer aided engineering design and manufacturing: a fourth industrial revolution perspective. Springer, Cham, pp 509–522

8. Khan WA et al (2025) Computer-aided numerical analysis techniques. In: Computer aided engineering design and manufacturing: a fourth industrial revolution perspective. Springer, Cham, pp 305–355
9. Khan WA et al (2025) Engineering design. In: Computer aided engineering design and manufacturing: a fourth industrial revolution perspective. Springer, Cham, pp 11–63
10. Hammad M, Abbas G, Khan WA (2024) IMS framework for N11 and D8 countries. In: Integration of heterogeneous manufacturing machinery in cells and systems. CRC Press, Boca Raton, pp 15–32
11. Crnobrnja J et al (2024) Digital transformation towards human-centricity: a systematic literature review. In: IFIP International Conference on advances in production management systems. Springer, Cham
12. Hammad M et al (2025) Lightweight and robust key agreement for securing IIoT-driven flexible manufacturing systems. IEEE Internet Things J
13. Abuhasel KA (2023) A linear probabilistic resilience model for securing critical infrastructure in Industry 5.0. IEEE Access 11:80863–80873
14. Mingaleva Z, Startseva O, Gerasimov A (2023) Aligning the environmental and social values of the sustainable development concepts, Industry 5.0 and Society 5.0. In: E3S Web of conferences, vol 460. EDP Sciences
15. Hammad M et al (2025) Smart manufacturing blueprint: navigating Industry 4.0 across diverse sectors. Springer
16. Rame R, Purwanto P, Sudarno S (2024) Industry 5.0 and sustainability: an overview of emerging trends and challenges for a green future. Innov Green Develop 3(4):100173
17. Alves J, Lima TM, Gaspar PD (2023) Is Industry 5.0 a human-centred approach? A systematic review. Processes 11(1):193
18. Hammad M et al (2023) A provable secure and efficient authentication framework for smart manufacturing industry. IEEE Access 11:67626–67639
19. Khan WA et al (2025) Remanufacturing. In: Computer aided engineering design and manufacturing: a fourth industrial revolution perspective. Springer, Cham, pp 187–211
20. Khan WA et al (2025) Future manufacturing techniques. In: Computer aided engineering design and manufacturing: a fourth industrial revolution perspective. Springer, Cham, pp 459–480
21. Khan WA et al (2025) Standards for CAD/CAE/CAM. In: Computer aided engineering design and manufacturing: a fourth industrial revolution perspective. Springer, Cham, pp 369–395
22. Golovianko M et al (2023) Industry 4.0 vs. Industry 5.0: co-existence, transition, or a hybrid. Procedia Comput Sci 217:102–113.
23. Khan WA et al (2025) Mathematics for CAD/CAE/CAM. In: Computer aided engineering design and manufacturing: a fourth industrial revolution perspective. Springer, Cham, pp 357–368
24. Khan WA et al (2025) Use of databases for CAD/CAE/CAM/CIM. In: Computer aided engineering design and manufacturing: a fourth industrial revolution perspective. Springer, Cham, pp 397–407
25. Khan WA et al (2025) Communication protocols—legacy and Industry 4.0. In: Computer aided engineering design and manufacturing: a fourth industrial revolution perspective. Springer, Cham, pp 485–504
26. Khan WA et al (2025) Nexus of CIM, Industry 4.0, Industry 5.0, and quantum computing. Computer aided engineering design and manufacturing: a fourth industrial revolution perspective. Springer, Cham, pp 551–557
27. Khan WA et al (2025) Quantum computing. In: Computer aided engineering design and manufacturing: a fourth industrial revolution perspective. Springer, Cham, pp 537–549
28. Dewantara H, Hafish M, Aldi Pranata M (2024) The impact of innovation strategies on competitive advantage: the role of students in the digital creative industry during Industry 5.0. J Econ Entrepreneur Manage Bus Account 2(2):77–83
29. Hammad M, Islam MS, Salam MA, Jafry AT, Ali I, Khan WA (2023) Framework for the implementation of smart manufacturing systems: a case in point. Processes 11(5):1436

30. Callari TC et al (2024) Shaping the future industry 5.0: ethical and societal implications of artificial intelligence and collaborative robotics technologies in entrepreneurship. Studi organizzativi: XXVI(2):210–236

Chapter 2
Historical Context and Evolution of Industry 5.0

2.1 Objectives

- To undertake a critical overview of the important changes in technology and society in the four great industrial revolutions.
- To evaluate the advantages and drawbacks of Industry 4.0, speaking in terms of automation and digitalization and their impact on society.
- To discuss the development of Industry 5.0 as the reaction to the Industry 4.0 issues, especially concerning human–machine synergy, sustainability and ethical responsibility.
- To critically examine the effects of each of the industrial revolutions on the labor market, and the environment.
- To emphasize the major technological achievements, namely, AI, robotics, and IoT, which have already influenced and will keep influencing the development of any industry in the future.

2.2 Review of the Past Industrial Revolutions: Major Technology and Society Transformations

The industrial revolutions have been a revolutionizing factor to the modern world. They have transformed economic systems, social patterns and the technological environment and all of them have innovative solutions that have changed the history of the world. There have been four great revolutions in industry and each one started radical change in society and in technology [1].

M. Hammad et al., *Industry 5.0: The Human-Centric Future of Industry*, Springer Series in Advanced Manufacturing, https://doi.org/10.1007/978-3-032-12622-1_2

2.2.1 *First Industrial Revolution (Late Eighteenth to Early Nineteenth Century): Mechanization Uprising*

- **Technological Change**: The First Industrial Revolution was stimulated by the advent of mechanized production and popularization of steam engines, which increased productivity to a tremendous degree. Important inventions like spinning jenny, power loom and steam locomotive transformed such industries as textile industry and transportation.

Spinning Jenny: It is a multi-spindle spinning frame invented by James Hargreaves in 1764 and it increased the production of the textile by a large margin.

Steam Engine: Steam engine was perfected by James Watt in late 18 th century and it was used to power machines, ships and trains and thus became the backbone of industrialization.

Such technological innovations preconditioned mass production and formation of industrial economies.

- **Societal Impact**: The First Industrial Revolution transformed the agrarian economies into urbanized and industrialized economies. Rural dwellers migrated to the cities in search of factory jobs, and this gave rise to industrial cities and urbanization.

Labor Shift: Labor in factories had increased greatly, and it was long hours, low pay, and even unsafe working conditions.

Social Problems: Cities were overcrowded and this caused unhealthy lifestyles and the sanitary, health and housing problems could not keep with the number of people in the city. The new working class in the industries was subjected to harsh conditions and this led to the onset of the labor movements.

Although this was the time of economic boom, there was also a massive increase in social inequality whereby most of the wealth was concentrated on industrialists and landowners [2].

2.2.2 *Second Industrial Revolution (The End of the Nineteenth Century to the Beginning of the Twentieth Century): Mass Production and Electricity*

- **Technological Change**: The Second Industrial Revolution was accompanied by inventions that formed the basis of contemporary mass production and electric power. Important inventions were the assembly line, electricity and the internal combustion engine.

Assembly Line: Assembly line was introduced by Henry Ford in the automotive industry and enabled efficient and mass production of goods, especially automobiles. This cut the manufacturing cost drastically and the products became affordable to the masses.

Electricity: With the introduction of electrical power, factories could be operated continuously, and this made them more productive and with the provision of good working conditions (e.g., lighting and heating).

Internal Combustion Engine: The invention of engine driven vehicles which transformed the way transportation and mobility was done [3].

- **Societal Impact**: The Second Industrial Revolution changed the economies by the use of mass production methods where bulky production was produced at cheap prices. Consumerism was launched and new products were available at a cheaper price.

Industrial Giants: Businesses became bigger and attained the status of industrial giants with international presence. Big companies in the steel, chemical and auto industries were formed.

Class Divide: Mass production not only added wealth, but it also contributed to the growing levels of division between classes. There were poor wages and working conditions that had resulted in great labor unrest among the working class. The widening gap between the industrial capitalists and the working people was one of the key issues of that period.

Global Trade: Global trade networks expanded as a result of the creation of railways, ships and the telegraph, and increased the scope of the industrialized countries [4].

2.2.3 Third Industrial Revolution (Late Twentieth Century) Digital Revolution and Automation

- **Technological Change**: The third industrial revolution, which is also referred to as the digital revolution, was instigated by the developments in computing, automation and information technology. Microelectronics, robotics, and personal computing were technologies that brought changes in industries especially in manufacturing and communication.

Computers and Automation: Automation was introduced through computers and robotics in the manufacturing system thus eliminating the need for manual labour and increasing productivity.

Internet and Personal Computing: The invention of the Internet and the spread of personal computers led to digital technology to the people and linked both people and businesses across the globe.

Digital revolution came with a new way of working including online communication, digital services and e-commerce [5].

- **Societal Impact**: The Third Industrial Revolution changed the economies into a knowledge-based economy where information was a valuable asset. There was more need for skilled labor especially in the IT and engineering sector.

Labor Market Shift: With the automation of most manual jobs, the demand of manual labor force declined, and the labor market started to value highly skilled labor to design, manage and program automated systems.

Global Connectivity: The emergence of the Internet enabled the possibility to communicate globally, and the world got to be more connected and promoted the development of the digital economy.

Problems: Although the digital revolution increased productivity, it also caused job loss and wealth disparity as most low-skill employment opportunities were taken over by robots. The digital divide was also evident as a section of the communities and countries did not have access to new technologies [6].

2.2.4 Fourth Industrial Revolution (Twenty-First Century): Cyber-Physical Systems and Digitalization

- **Technological Change**: Cyber-physical systems, IoT, AI, big data, and advanced robotics are the characteristics of the Fourth Industrial Revolution, also known as Industry 4.0. These technologies will allow the development of intelligent factories, the interaction of machines and systems, autonomous communication and optimization of processes in real-time [7].

Cyber-Physical Systems: These are systems that combine physical process with computer-based algorithms to produce smarter and responsive systems.

IoT and AI: The common usage of IoT devices and AI has allowed collecting and analyzing data in real-time, making decisions based on the insight obtained with the help of connected devices.

Blockchain and Cloud Computing: Blockchain enhances transparency and security in every industry, whereas cloud computing allows storing and processing data on a scale [8].

- **Societal Impact**: The fourth industrial revolution has increased the rate of digitalization, and the industries are becoming more efficient and interconnected. Although it has resulted in significant advances of automation and customized products, it poses new challenges [9].

Customization/Personalisation: The consumers are now demanding a more personalized product, and the technologies allow the creation of customized items on a large scale. To provide custom experience to the customers, businesses are adopting AI and machine learning.

Workforce Transformation: A greater dependence on digital technologies due to industry 4.0 has led to the fact that the workforce has to be highly qualified and able to control and optimize automated processes. This change entails unending upskilling and reskilling.

Issues: The growing use of digital technologies has brought about the issue of cybersecurity, data privacy, and the possibility of increased inequality because some jobs will be gone or will be closed to less-skilled employees [10].

Figure 2.1 represents the four industrial revolutions from the First (steam engine), Second (electricity and factories), Third (automation), and Fourth (AI and digital technologies).

2.3 Lessons of Industry 4.0: Advantages and Drawbacks of Automation and Digitalization

The fourth industrial revolution, also known as the Industry 4.0, has revolutionized the manufacturing and business process where innovative technologies like automation, artificial intelligence (AI), big data, IoT, and cloud computing have been adopted. Although the positive effects of such technological developments are well recognized, Industry 4.0 has revealed some of these flaws, and Industry 5.0 is supposed to overcome them through the human-centric, adaptive, and ethical approaches. In this section, the main advantages and shortcomings of Industry 4.0 will be discussed, and the opportunity to learn from these experiences and develop Industry 5.0 will be reviewed [11].

2.3.1 Advantages of the Industry 4.0

- **Enhanced efficiency and productivity**: Automation, robotics and AI have had a huge effect of efficiency and productivity in business and production processes. Automation of menial processes allows the industries to manufacture goods in large quantities using less resources and less human labor.

Key Impact: Reduced production cycle, accurate operations and minimization of human error. Automation enables companies to produce at a high quality and at the same time produce more hence leading to cost reduction and high profit margins.

Example: Robots can be used in automotive manufacturing to do fine welding, painting and assembly work and this saves the cost of human labor, yet the quality of the products will be the same [12].

Fig. 2.1 The first to fourth industrial revolution, from steam engine to modern technology

- **Increased Data Usage**: Big data analytics and Internet of Things (IoT) have helped industries to collect and analyze massive data to have a better understanding of customer behavior, efficiency, and maintenance.

Key Impact: With the help of data-driven decisions, the companies will be able to streamline their operations, enhance their products, and even foresee certain problems like machine failure before it is even possible to experience it.

Case Study: Retailers can analyse purchasing trends of the customers using big data analytics and customise the marketing strategy leading to growth in sales and customer retention. Likewise, data-driven predictive maintenance enables the manufacturers to minimize the amount of downtime by carrying out maintenance before the equipment breaks down.

- **Customization and Personalization**: Industry 4.0 also made mass customization possible since businesses can now manufacture goods according to the preferences of each consumer on a large scale using digital technologies. This change will allow businesses to produce very customised products without compromising the efficiency of the production process.

Significant Impact: The consumers also get access to customized products that respond better to their needs whereas the businesses are capable of getting to greater heights of customer satisfaction and loyalty.

Example: Nike and Adidas employ 3D printing and digital designing technology to make custom-fitted shoes to athletes that provide a more personalized experience to the consumers.

- **Global Connectivity**: The focus on IoT and cloud computing in the industry 4.0 has helped to connect the world, allowing companies to control their supply chains, remote monitoring, and instant communication regardless of the borders.

Important Implication: The organizations are now able to track the activities in various locations, streamline the supply chain by tracking the inventory levels in real time and make faster decisions and more informed decisions in various markets around the world.

Example: Amazon has implemented cloud computing and IoT to coordinate its international supply chain, making it possible to monitor the stock in real-time and automate operations in warehouses. This has resulted in reduced delivery time and increased satisfaction with the customers [13].

2.3.2 Shortcomings of the Industry 4.0

However, along with the great opportunities, the high pace of automation and digitalization created a number of constraints that should be overcome in Industry 5.0.

- **Job Displacement**: Automation and the use of AI technology has displaced a lot of jobs, especially those of low-skill and labor-intensive nature. Automation has brought about efficiency but has raised questions about what is going to happen to the jobs of workers in the manufacturing, retail and logistics sectors [14].

Critical Impact: A large number of workers, particularly those that lack special skills, are likely to lose their jobs to machines, causing greater unemployment in some areas.

Example: Robots have taken the jobs of many human workers in the case of automotive manufacturing in assembly, welding and painting among others leaving less work on manual labor to be done by workers who lack technical knowledge.

- **Skill Gaps**: The need to have highly specialized technical skills has increased tremendously due to use of advanced automation and digitization of industries. Nonetheless, most employees have not been trained to work in such new environments and therefore there is a mismatch in skills and adaptation of the workforce.

Key Impact: Industries are struggling to find employees with the necessary STEM skills (science, technology, engineering, and mathematics) due to the spread of automation and digitalization, leaving a lot of employees behind.

Example: With the introduction of robotics and AI-based systems in the factories, the workers are expected to possess some technical skills to operate and work on the systems. But the problem is that most of the employees in the traditional manufacturing sectors do not have the required skills to take up these new jobs.

- **Data Privacy and cybersecurity**: The spread of networked devices and exchange of huge data have greatly raised the vulnerabilities of cybersecurity and risk of data privacy. The greater the integration of systems, the higher the chances of cyber-attacks, data leakages and unauthorized access to confidential information.

Key Impact: In the more interconnected world, businesses have to deal with the challenge of maintaining the security and privacy of the customer data, intellectual property, and other sensitive business information.

Example: In financial services, the combination of digital banking and the connection of devices through the Internet of Things has turned such institutions into easy targets of cybercriminals. Security breaches may result in loss of data, loss of money and loss of reputation of a company [15].

- **Over dependence on Technology**: Although this has enhanced efficiency, the over-reliance on automated systems also exposes industries to the risk of system failure and technological hitch. Excessive use of technology in the business processes may cause serious interruption in case of system failure or bugs.

Key Impact: A malfunction in any single component of the automated system may lead to overall disturbances, suspension of the production, postponement of the services, or logistical breakdowns.

Example: In airlines, the use of automated ticketing, check-in and flight management systems may result in critical operational failures due to system breakdown, and this may affect the customer satisfaction and revenues of the company [16].

Figure 2.2 highlights the advantages of industrial advancements like increased productivity and global connectivity, and drawbacks such as job displacement, skill gaps and excessive use of technology.

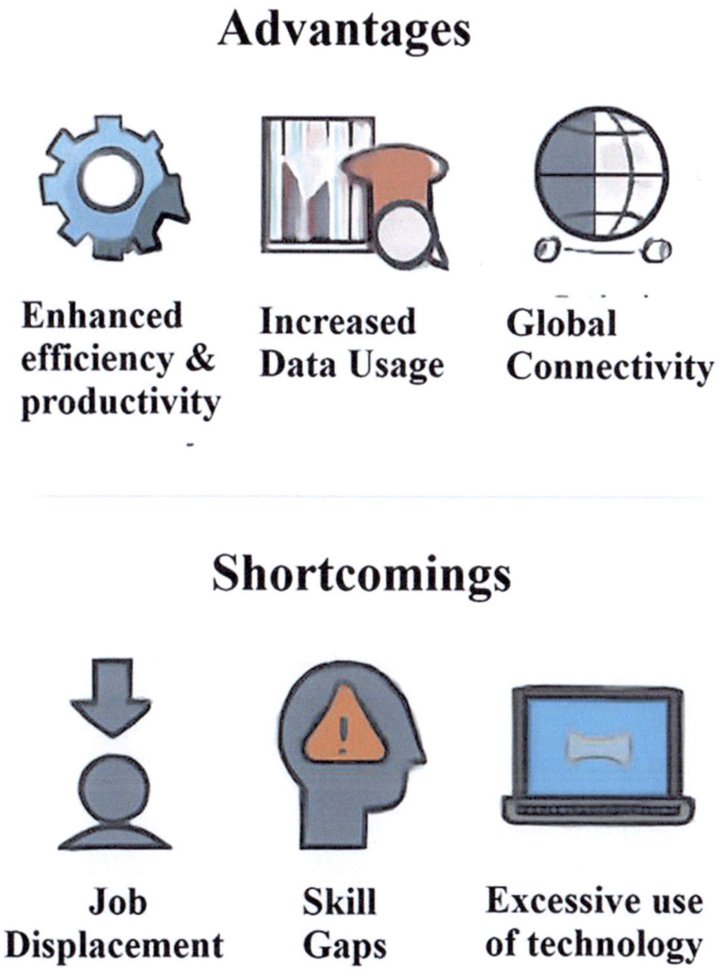

Fig. 2.2 Benefits and challenges of industrial advancements

2.4 Introduction of the Industry 5.0

Industry 5.0 is aimed at overcoming the deficiencies of Industry 4.0 with the emphasis on human-related solutions and adaptive, sustainable activities. Although the major benefits of automation and digitalization have been achieved, Industry 5.0 is supposed to improve the cooperation of people and machines, making sure that employees are in the center of innovation but using technology to boost creativity, efficiency, and ethical activity.

Industry 5.0 focuses on upskilling, reskilling, the cooperation of humans and machines, which helps employees to perform more strategic and creative tasks and work with high-tech systems without difficulties. Industry 5.0 promotes a lifelong learning and training system, which makes sure that employees acquire not only technical competencies but also human-related skills to collaborate with the new technologies. Industry 5.0 will encourage strong cybersecurity systems and data privacy policies to curb cyber-attacks and secure confidential data. Industry 5.0 promotes resilient systems where human supervision is integrated with automated technology so that human judgment and decision-making are at the center of the critical operations [17].

2.4.1 The Emergence of Industry 5.0

The direct answer to the shortcomings of Industry 4.0 is the emergence of Industry 5.0. Although Industry 4.0 was defined by the high rate of adoption of automation and efficiency-based processes, it failed to consider the human element in the production process and paid much attention to efficiency on the part of machines. Industry 5.0 will aim to bring back human workers into the manufacturing and production process where the focus will be more on human–machine synergy, sustainability, and industrial system resilience. It is focused on industrial processes to be more humanistic, innovative, and environmentally friendly.

- **Human–Machine Synergy**: Industry 5.0 is all about human–machine synergy, and the advanced technologies, such as collaborative robots (cobots), artificial intelligence (AI), and automation, are combined with human abilities to supplement and complement them, not to substitute them. The collaboration between people and machines does not only enhance productivity but also enhances creativity and innovation in the workforce.

Collaborative Robots (Cobots): The creation of the so-called cobots, or collaborative robots, is one of the most important technologies that make it possible to achieve human–machine synergy in Industry 5.0. In contrast to the usual industrial robots, which in most cases substitute human efforts, cobots are made to collaborate with human employees, improving productivity and creativity. Cobots perform repetitive, hazardous or dull jobs, e.g. on the assembly line, in material handling or in hazardous processes, so that the human worker can concentrate on more valuable work that needs critical thinking, problem-solving, and emotional intelligence.

Case study: Cobots may be used in the automotive sector to help in the assembly of complicated components, including engine components, which will relieve the physical burden of the workers and speed up the assembly process and accuracy, whereas human can be used in the assembly of the complex parts that need creativity and management [18].

- **Creative Collaboration**: Although AI and automation systems can simplify the manufacturing process and make it more efficient, the human mind, creativity, judgment, and decision-making are still crucial to the industries that require continuous innovation and personalization. Industry 5.0 focuses on the partnership between humans and AI in order to develop products that possess human knowledge, creativity and ethical accountability [19].

Human Creativity: Although AI would help in some areas like data analysis, optimization, and automation, the human workers are incorporated in the design and development process, where their creativity and intuition could be exploited to come up with innovative products that meet individual needs.

Examples: In fashion or product design, AI may create a model or a prototype according to the data, yet human designers will have to give it some creativity, aesthetics, and cultural relevance. The synergy can give it the opportunity to create both functional and emotionally appealing products to the consumers [20].

- **Sustainable Development**: One of the missing links in Industry 4.0 was the emphasis on sustainability of the environment. Although automation has had immense efficiency, it has in most cases led to over production, depletion of resources and generation of wasteful materials. Industry 5.0 focuses on the concept of sustainability, implementing practices that focus on the circular economy, green manufacturing, and the minimization of the environmental impact.

Circular Economy: Industry 5.0 incorporates the concept of the circular economy, in which products are made to be reusable, remanufactured and recycled, and to produce a minimum amount of waste and to have minimal environmental impact. Instead of operating on the linear economy model of take, make, dispose, Industry 5.0 encourages the closed-loop model where resources are reused.

Automation in Circular Economy: Automation and AI are important in maximizing resource utilization and waste management, among other things, making sure that products are made to be easily disassembled, recycled, and remanufactured at the end of their life cycle.

Example: In electronics Industry, Industry 5.0 promotes the modular design and recyclable materials in product development to make sure that the old devices can be easily refurbished or repurposed, decreasing the e-waste and being sustainable.

- **Green Manufacturing**: In Industry 5.0, the automation and AI-driven systems are applied to track the energy consumption, optimise resources, and minimize emissions to make the manufacturing process more environmentally friendly. Sustainability is not a regulatory compliance issue, but it is part of the core manufacturing where technology can be used to have friendly environmental manufacturing processes.

Energy Optimization: AI and IoT devices can be used to monitor energy consumption in real-time and thus enable manufacturers to optimize energy consumption, minimize unwarranted wastage and cut carbon footprints.

Example: In the field of textile manufacturing, AI tools can monitor the use of water and energy and waste production, which will enable manufacturers to streamline their operations in order to minimize environmental harm without losing their productivity [21].

Figure 2.3 shows human–machine synergy, creative collaboration, sustainable development, and green manufacturing practices.

2.4.2 Filling the Gaps: The Solutions that Industry 5.0 Give to Industry 4.0

Industry 5.0 works as the direct response to the shortcomings of Industry 4.0, especially its over-automation and efficiency without regard to human creativity, ethical responsibility, and sustainability.

i. **Human-Centered Technology**: Industry 5.0 addresses the vacuum left by Industry 4.0, which is based on machine-driven automation by bringing the human factor back to the production process. This is not meant to substitute human workers but rather to augment their abilities by letting them work with intelligent systems. Industry 5.0 is based on human–machine interaction with the understanding that humans are the ones who introduce creativity, moral decision-making, and emotional intelligence into the leading role in industrial processes.

Ethical AI: Industry 5.0 also focuses on creating ethical AI whereby machine judgment reflects the human beliefs and the needs of society. This alleviates the fears of automation and loss of jobs so that technology is used to benefit humankind and not to substitute for it.

ii. **Eco-friendly Industries**: Industry 5.0 is an answer to the environmental issues which Industry 4.0 has revealed, and the industry is more concerned with creating sustainable production systems that will optimize economic and environmental performance. Such technologies as AI, IoT, and blockchain will be critical in making sure that the industrial processes are more resource-efficient, less wasteful, and have fewer carbon footprints.

Fig. 2.3 The future of manufacturing with human–machine collaboration

Resource Efficiency: Industry 5.0 focuses on closed loop systems that will reuse resources, manufacture customized products out of sustainable material and reduce emissions by monitoring in real time.

iii. **Eco-friendly Industries**: Industry 5.0 is an answer to the environmental issues which Industry 4.0 has revealed, and the industry is more concerned with creating sustainable production systems that will optimize economic and environmental performance. Such technologies as AI, IoT, and blockchain will be critical in making sure that the industrial processes are more resource-efficient, less wasteful, and have fewer carbon footprints.

Workforce Inclusivity: Although Industry 4.0 brought about skill gaps and widened inequality, Industry 5.0 is all about workforce inclusivity; it not only trains workers on technical skills but also on the ability to work collaboratively to solve problems and make decisions creatively. Upskilling and reskilling are also a part of Industry 5.0 as they guarantee that every worker will be able to succeed in the environment where the collaboration between humans and machines will be the primary factor.

Industry 5.0 is able to address the most important gaps left behind by the Industry 4.0 based on human–machine synergy, sustainable development, and inclusive innovation. It aims at establishing an equal and strong industrial ecosystem, in which the advantages of humanity and robotics are used to increase productivity, creativity, and sustainability. Through the reintroduction of human creativity and responsibility into it, Industry 5.0 offers to develop an ethically responsible, socially inclusive, and environmentally sustainable industrial environment [22].

2.5 Comparative Study of the Effects of Industrial Revolution on Labor Markets and Environment

Industrial revolutions have drastically changed not only the technological scenery but also work markets as well as environmental conditions. All revolutions have presented unique outcomes to the labor force and the earth with technology causing major changes. The major differences in the industrial revolutions as far as labor markets and the environment are concerned are discussed below.

2.5.1 Effects on Labor Market

Every industrial revolution has seen the labor market change drastically due to the influence of technology as well as the change in the means of production. These revolutions have brought in new jobs, and this has caused job loss and social transformation.

- **First Industrial Revolution**: The shift of workers in the rural agricultural sector into urban-based factory-based labor took place during the First Industrial Revolution. This wave of migration was a major transition in the trend of employment, whereby people moved out of the agricultural sector based on land into industrialization which was factory-based.

Working Conditions: Factories, which were usually long hours, low wages and bad working conditions were a big change. There was child labor and exploitative and hazardous conditions that were subjected to the workers which led to the emergence of labor movements that promoted rights of workers.

Mass Employment: Although industrialization led to huge employment opportunities in the factories, these were mostly menial and physically stressing. The transition of production to the factory environment established mass employment, at the expense of the welfare of workers, which included exploitation, unsafe working conditions and job security.

Example: In textile production, the use of such devices as the spinning jenny and power loom decreased the demand on the traditional hand-weaving labor but instead gave rise to a new need of low-wage, factory work [23].

- **Second Industrial Revolution**:

The Demand of Skilled Labor: The Second Industrial Revolution also led to mass production particularly in the steel, chemicals and electricity industries. This era needed labor with skills to work in new industries, especially to run heavy machines and to work on assembly lines.

Stiff Social Classes: This revolution also led to the establishment of stiff social classes where the industrialists were on top, the skilled craftsmen in the middle and factory workers at the bottom. Although factory work was a source of employment, it was characterized by a big gap between highly skilled workers and those who worked on lower wages.

Social Unrest: With the growth in production and economic growth, there was also the emergence of industrial giants and with this came the labor unrest and the division of classes as the workers were also exploited through the provision of long hours of work, low payment and poor working conditions.

Case in point: The assembly line invented by Henry Ford changed the way automobiles were manufactured as it gave skilled workers an opportunity to earn their living in new automotive plants. This also resulted in the emergence of a working industrial workforce, which was frequently exposed to harsh conditions, and was permanent [24].

- **Third industrial Revolution**:

Automation and Displacement: The Third Industrial Revolution was the start of automation whereby manual work in the manufacturing industry was being substituted by machines and computers. Traditional jobs (manufacturing jobs) were lost, and new jobs were created in information technology (IT), programming and data analysis.

Knowledge-Based Economy: The revolution transformed the economies into knowledge-based economy where the need of skilled workers in the field of IT, data management and engineering is increased. It opened up new opportunities as well as caused polarization of the jobs with high skill jobs thriving and the low skill manufacturing jobs dying.

Worldwide Labor Transformation: As the world became more globalized and digital, there were more demands of highly skilled professionals particularly in the tech-driven sectors, and low-skill workers were faced with the problem of job displacement.

Example: The appearance of personal computers and automation in production caused displacement of jobs in the old industries but established new types of jobs in software development, network management and technical support [25].

- **Fourth industrial revolution**:

Enhanced Automation and Robotics: During the Fourth Industrial Revolution, robotics, artificial intelligence, and machine learning continued to replace the manual workers especially in routine and repetitive tasks. It also brought new workplaces in the sphere of AI, data science, and cybersecurity.

Skill Gap and Digital Divide: On the one hand, the emergence of new opportunities led to the polarization of the labor market, as the gap between the technical and non-technical population increased. The dependence on high-tech skills created more disparity in the workforce owing to the fact that manual and low-skill jobs were no longer in demand.

Future of Work: Automation kept on changing and there was increased emphasis on the upskilling and reskilling of the workforce. Industry 4.0 required that the employees learn new technical skills to succeed in a world where technology, AI, and automation were on the front burner.

Example: In the manufacturing industry, the implementation of robots and AI caused the loss of jobs, but reskilling programs were introduced to help the workers take up jobs in the technology sector such as working on robots, AI software, and data analysis [26].

- **Industry 5.0**:

Human–Machine Collaboration: Industry 5.0 aims at reconsidering human creativity and emotional intelligence as the key elements of the workforce. It does not mean replacing human beings with machines but focuses on human–machine synergy in which employees will be working alongside robots and AI to increase innovation and productivity.

Reskilling and Creativity: Industry 5.0 proposes reskilling of the workforce and creation of jobs that are centered around innovation, creativity and problem solving. It focuses on the improvement of human abilities by working with intelligent systems, and the result is a more balanced overview of work distribution.

Example: The use of AI and cobots in the creative industries can aid in such tasks as design, whereas human labor can bring its creativity, sense of aestheticism, and emotional intelligence to create novel products that cannot be developed by machines only [27].

2.5.2 Environmental Impacts

The extent of the effect on the environment of each of the industrial revolutions has been different and the older revolutions have played a major role in pollution, depletion of resources and degradation of the environment. Industry 5.0, in its turn, makes sustainability and environmentally friendly production the point of focus.

- **First Industrial Revolution**:

Coal Powered Production: The first industrial revolution used coal as fuel which had serious environmental impacts like air pollution, deforestation and depletion of natural resources. The period was dominated by economic development which in most cases sacrificed environmental health.

Pollution: The fact that steam engines and coal-powered factories were being used in great numbers led to a huge rise in emissions and pollution of the environment, and this was one of the first steps toward industrial pollution.

Example: The increase of textile factories and the growth of railroads with the use of coal-powered steam engines led to huge pollution of the cities and the exhaustion of natural resources [28].

- **Second Industrial Revolution**:

Fossil Fuels and Pollution: Fossil fuels were used during the Second Industrial Revolution, and the use of electricity and internal combustion engines, even more, added to their dependence on fossil fuels. This era caused further air pollution, water pollution and the preliminary symptoms of worldwide environmental issues.

Resource Depletion: Mass production and the ever-growing demand for steel, chemicals and other industrial products led to the overexploitation of the natural resources, which posed a long-term problem of sustainability.

Example: The emergence of steel industry and the emergence of automobiles manufacturing industries raised the demand of coal which has not only led to environmental degradation but also to a lot of carbon emission.

- **Third Industrial Revolution**:

Digital Efficiency Versus E-Waste: The Third Industrial Revolution brought us energy efficient technologies, but the creation of digital devices and electronics, as well as data centers, created new environmental problems, including e-waste and energy-intensive processes.

Data Centers and Energy: The growing demand for data storage and the growth of cloud computing led to consumption of more energy. Due to this fact, data centers became one of the large consumers of electricity and this leads to concerns about the environmental impact of digital infrastructures.

Case in point: The worldwide proliferation of smart phones and personal computers has increased e-waste and the data centers have been contributing to the increased amount of energy being consumed and this calls out that there is a need to produce more environmentally friendly technology [29].

- **Fourth Industrial Revolution**:

Energy Optimization and Sustainability: Energy optimization and sustainability can be achieved with the help of smart technologies (AI, IoT, cloud computing) as a part of Industry 4.0, allowing it to optimize energy consumption, eliminate waste, and increase resource efficiency. Nonetheless, the extensive use of digital technologies and mining rare-earth metals to make electronics are threats to the environment.

Data Centers and Resource Utilization: With increased migration of industries to cloud computing, the issue of the environmental impact of large-scale data centers has not diminished since they are consuming a lot of energy.

Use case: Smart cities will be able to use IoT and AI to maximize energy consumption, minimize congestion, and minimize emissions. Nevertheless, the environmental implication of running data centers to sustain such technologies is of concern.

- **Industry 5.0**:

Circular Economy: industry 5.0 incorporates the principles of circular economy, which emphasise reuse, remanufacturing and recycling to make sure that the waste is minimal and that the environmental impact is minimized. This is to develop systems that reuse resources continuously, minimize waste and recycle materials during production.

Green Manufacturing: AI and automation of Industry 5.0 will allow the use of eco-friendly production processes, so that the focus of industrial activities is on sustainability. The purpose of Industry 5.0 is to overcome the environmental problems that the previous revolutions have caused by making systems more energy efficient and low emission [30].

Example: The use of AI-powered energy control and waste management in the manufacturing environment will also assist to lower carbon footprints, consume resources, and decrease environmental degradation.

Industrial revolutions have significantly and in some cases affected labor and the environment. The fifth Industrial Revolution (Industry 5.0) offers to strike a balance between technical progress and human-centered work and sustainability and fill in the gaps created by the previous revolutions [31–48]. Industry 5.0 aims at establishing a more balanced job market and alleviating the environmental issues affecting the previous generation through human–machine cooperation and green manufacturing processes integration.

2.6 Learning Outcomes

Upon completing this chapter, readers will be able to:

- Identify and describe the key technological changes that define each industrial revolution.
- Analyze the societal and economic impacts of each industrial revolution, particularly in terms of labor markets and environmental consequences.
- Evaluate the benefits and limitations of Industry 4.0 and explain how these challenges are addressed by the advent of Industry 5.0.
- Understand the role of human creativity and ethical considerations in Industry 5.0's approach to technological integration.
- Discuss the concept of a circular economy and its importance in the sustainable practices promoted by Industry 5.0.

2.7 Review Questions

1. How did the societal impact of the First Industrial Revolution influence the subsequent revolutions?
2. In what ways did the technological advancements of the Second Industrial Revolution contribute to the rise of global trade networks and industrial giants?
3 Discuss the role of automation and digital technologies in the Third Industrial Revolution. How did the shift towards a knowledge-based economy impact job markets, and what challenges did it pose in terms of income inequality and labor displacement?
4. How does Industry 5.0 address the gaps left by Industry 4.0 in terms of ethical responsibility, job creation, and environmental sustainability?
5. Provide examples of how human–machine synergy can be implemented in modern industries.
6. Considering the environmental impacts of each industrial revolution, how does Industry 5.0 prioritize sustainability?
7. Discuss the integration of green manufacturing and circular economy principles in mitigating the environmental consequences of industrialization.
8. How does the concept of human–machine collaboration in Industry 5.0 differ from the automation-driven approach of Industry 4.0?
9. Examine the labor market implications of Industry 5.0. What roles will emerge as a result of human–machine collaboration, and how can workers be reskilled to adapt to these new opportunities?
10. Compare the environmental concerns raised by each industrial revolution. How does Industry 5.0's focus on sustainability and resource conservation contrast with the environmental degradation seen during previous industrial revolutions?
11. In terms of technological advancements, how does Industry 5.0 integrate AI and automation into sustainable practices?

12. Provide an analysis of how AI-driven systems and collaborative robots can enhance eco-friendly production methods while reducing industrial emissions.

Appendix: Solution Manual

1. **How did the societal impact of the First Industrial Revolution influence the subsequent revolutions?**

   Solution: The First Industrial Revolution led to a massive rural-to-urban migration, driven by the rise of factory-based labor. The social shifts created urban overcrowding, poor working conditions, and the exploitation of labor, including child labor. These challenges highlighted the need for labor reforms and set the stage for the demand for more organized industrial workforces in subsequent revolutions.
2. **In what ways did the technological advancements of the Second Industrial Revolution contribute to the rise of global trade networks and industrial giants?**

   Solution: The Second Industrial Revolution's technological advancements in electricity and mass production methods, such as the assembly line, created an environment where goods could be produced at scale, thus making them more affordable. This contributed to a rise in consumerism and the formation of large industrial corporations. However, it also widened social class divides, creating tensions between factory workers and industrialists.
3. **Discuss the role of automation and digital technologies in the Third Industrial Revolution. How did the shift towards a knowledge-based economy impact job markets, and what challenges did it pose in terms of income inequality and labor displacement?**

   Solution: The Third Industrial Revolution's emphasis on digital technology and automation led to a knowledge-based economy, where industries increasingly required skilled workers in IT, programming, and data analysis. This shift displaced many low-skilled manufacturing jobs, exacerbating income inequality. However, the revolution also created high-skilled job opportunities, which led to job polarization.
4. **How does Industry 5.0 address the gaps left by Industry 4.0 in terms of ethical responsibility, job creation, and environmental sustainability?**

   Solution: Industry 5.0 addresses the ethical challenges of Industry 4.0 by emphasizing human-machine collaboration, ensuring that humans retain decision-making power and creativity. It promotes ethical AI development and reskilling of the workforce to meet new roles in innovation and sustainability.
5. **Provide examples of how human-machine synergy can be implemented in modern industries.**

Solution: For example, AI-powered cobots assist humans in creative tasks rather than replacing them, fostering a more human-centric approach to production.

6. **Considering the environmental impacts of each industrial revolution, how does Industry 5.0 prioritize sustainability?**

   Solution: Industry 5.0 places a strong emphasis on sustainability, integrating green technologies like renewable energy and recycling systems into production. Unlike previous revolutions, which prioritized economic growth at the expense of the environment, Industry 5.0 aims to minimize waste, conserve resources, and reduce carbon emissions through eco-friendly practices and circular economy principles.

7. **Discuss the integration of green manufacturing and circular economy principles in mitigating the environmental consequences of industrialization.**

   Solution: Industry 5.0 places a strong emphasis on sustainability, integrating green technologies like renewable energy and recycling systems into production. Industry 5.0 aims to minimize waste, conserve resources, and reduce carbon emissions through eco-friendly practices and circular economy principles.

8. **How does the concept of human-machine collaboration in Industry 5.0 differ from the automation-driven approach of Industry 4.0?**

   Solution: Industry 5.0's human-machine collaboration contrasts with Industry 4.0's automation-driven model by reintroducing human creativity and problem-solving into the production process. Cobots and AI systems assist rather than replace workers, allowing them to focus on higher-level tasks. This paradigm benefits labor markets by creating new job opportunities centered around creativity, strategy, and innovation.

9. **Examine the labor market implications of Industry 5.0. What roles will emerge as a result of human-machine collaboration, and how can workers be reskilled to adapt to these new opportunities?**

   Solution: Industry 5.0 introduces new roles focused on creative collaboration and strategic decision-making. These roles require workers to be reskilled in areas such as AI management, robot maintenance, and sustainable production practices. Reskilling initiatives will ensure that workers can adapt to this new technological environment, creating a more diverse and equitable workforce.

10. **Compare the environmental concerns raised by each industrial revolution. How does Industry 5.0's focus on sustainability and resource conservation contrast with the environmental degradation seen during previous industrial revolutions?**

    Solution: The environmental degradation caused by previous industrial revolutions was primarily driven by the use of fossil fuels and mass production methods. Industry 5.0 contrasts with this by prioritizing sustainability, with green manufacturing practices, energy-efficient systems, and a circular economy that focuses on reuse, recycling, and reducing waste throughout the product lifecycle.

11. **In terms of technological advancements, how does Industry 5.0 integrate AI and automation into sustainable practices?**

    Solution: Industry 5.0 integrates AI and automation into sustainable practices by using AI-driven systems to optimize energy consumption, reduce waste, and monitor resource usage.
12. **Provide an analysis of how AI-driven systems and collaborative robots can enhance eco-friendly production methods while reducing industrial emissions**.

    Solution: Cobots can handle tasks such as recycling and energy management, ensuring that the benefits of automation do not come at the cost of environmental responsibility. These advancements help reduce emissions and promote eco-friendly production methods.

Bibliography

1. Melnyk LH et al (2019) The effect of industrial revolutions on the transformation of social and economic systems
2. Cowan RS (1976) The" industrial revolution" in the home: household technology and social change in the 20th century. Technol Cult 17(1):1–23
3. Hammad M et al (2023) Framework for the implementation of smart manufacturing systems: a case in point. Processes 11(5):1436
4. Taalbi J (2019) Origins and pathways of innovation in the third industrial revolution. Ind Corp Chang 28(5):1125–1148
5. Mohajan H (2021) Third industrial revolution brings global development, pp 239–251
6. Adamik A (2019) Creating a competitive advantage in the age of Industry 4.0. Problemy Zarządzania 172(82):13–31
7. Liao Y et al (2018) The impact of the fourth industrial revolution: a cross-country/region comparison. Production 28:e20180061
8. Malik A (2019) Creating competitive advantage through source basic capital strategic humanity in the industrial age 4.0. Int Res J Adv Engineering and Science 4(1):209–215
9. Bonilla SH et al (2018) Industry 4.0 and sustainability implications: a scenario-based analysis of the impacts and challenges. Sustainability 10(10) (2018): 3740.
10. Brozzi R et al (2020) The advantages of Industry 4.0 applications for sustainability: results from a sample of manufacturing companies. Sustainability 12(9):3647
11. Dallasega P, Marengo E, Revolti A (2021) Strengths and shortcomings of methodologies for production planning and control of construction projects: a systematic literature review and future perspectives. Prod Plan Control 32(4):257–282
12. Mahnkopf B (2019) The '4th wave of industrial revolution'—a promise blind to social consequences, power and ecological impact in the era of 'digital capitalism'. EuroMemo Group (2019)
13. Enrique DV et al (2021) Advantages and difficulties of implementing Industry 4.0 technologies for labor flexibility. Procedia Comput Sci 181:347–352
14. Horvat D, Kroll H, Jäger A (2019) Researching the effects of automation and digitalization on manufacturing companies' productivity in the early stage of industry 4.0. Procedia Manuf 39:886–893

15. Hristoskova A et al (2021) An initial analysis of the shortcomings of conventional AI and the benefits of distributed AI approaches in industrial use cases. In: IFIP International conference on artificial intelligence applications and innovations. Springer, Cham
16. Holzinger A et al (2024) From Industry 5.0 to Forestry 5.0: bridging the gap with human-centered artificial intelligence. Curr Forest Rep 10(6):442–455
17. Hammad M et al (2025) Smart manufacturing in automotive industry. In: Smart manufacturing blueprint: navigating Industry 4.0 across diverse sectors. Springer, Cham, pp 47–85
18. Rame R, Purwanto P, Sudarno S (2024) Industry 5.0 and sustainability: an overview of emerging trends and challenges for a green future. Innov Green Develop 3(4):100173
19. Huang L, Jia Y (2022) Innovation and development of cultural and creative industries based on big data for industry 5.0. Sci Program 1:2490033
20. Li B, Song P (2022) Driving force mechanism of the core green technology innovation of equipment manufacturing enterprises towards industry 5.0 in China. Math Probl Eng 1:1404378
21. Hammad M et al (2025) Security challenges in smart manufacturing. In: Smart manufacturing blueprint: navigating Industry 4.0 across diverse sectors, pp 351–374
22. Dimanov DK (2024) Social and human costs of industrialization: a review of some negative effects on society during the first industrial revolution. Knowledge Int J 62(1):123–127
23. Milner B (2020) Essays on British labour markets during the Second Industrial Revolution. Dissertation, University of British Columbia
24. Alper AE, Alper FÖ (2020) Industry 4.0 revolution and its impacts on labor market. Çukurova Üniversitesi Sosyal Bilimler Enstitüsü Dergisi 29(3):441–460
25. Milošević IM et al (2022) Industry 4.0: limitation or benefit for success? Serb J Manage 17(1):85–98
26. Coelho P et al (2023) Industry 5.0: the arising of a concept. Procedia Comput Sci 217:1137–1144
27. Barca S (2011) Energy, property, and the industrial revolution narrative. Ecol Econ 70(7):1309–1315
28. Colombo AW et al (2017) Industrial cyberphysical systems: a backbone of the fourth industrial revolution. IEEE Indus Electron Mag 11(1):6–16
29. Leng J et al (2022) Industry 5.0: prospect and retrospect. J Manuf Syst 65:279–295
30. Matt DT et al (2020) Urban production–a socially sustainable factory concept to overcome shortcomings of qualified workers in smart SMEs. Comput Indus Eng 139:105384
31. Khan WA et al (2025) Industry 5.0. In: Computer aided engineering design and manufacturing: a fourth industrial revolution perspective. Springer, Cham, pp 523–535
32. Khan WA et al (2025) Computer aided manufacturing. In: Computer aided engineering design and manufacturing: a fourth industrial revolution perspective. Springer, Cham, pp 283–297
33. Khan WA et al (2025) Production planning and control. In: Computer aided engineering design and manufacturing: a fourth industrial revolution perspective. Springer, Cham, pp 425–453
34. Khan WA et al (2025) Manufacturing and production. In: Computer aided engineering design and manufacturing: a fourth industrial revolution perspective. Springer, Cham, pp 65–185
35. Hammad M et al (2025) Lightweight and robust key agreement for securing IIoT-driven flexible manufacturing systems. IEEE Internet Things J (2025)
36. Khan WA et al (2025) Computer-aided numerical analysis techniques. In: Computer aided engineering design and manufacturing: a fourth industrial revolution perspective. Springer, Cham, pp 305–355
37. Khan WA et al (2025) Engineering design. In: Computer aided engineering design and manufacturing: a fourth industrial revolution perspective. Springer, Cham, pp 11–63
38. Khan WA et al (2025) Pillars of Industry 4.0 and CAD/CAM. In: Computer aided engineering design and manufacturing: a fourth industrial revolution perspective. Springer, Cham, pp 509–522
39. Xu X et al (2021) Industry 4.0 and Industry 5.0—inception, conception and perception. J Manuf Syst 61:530–535
40. Hammad M et al (2023) A provable secure and efficient authentication framework for smart manufacturing industry. IEEE Access 11:67626–67639

41. Huang S et al (2022) Industry 5.0 and Society 5.0—comparison, complementation and co-evolution. J Manuf Syst 64:424–428
42. Khan WA et al (2025) Future manufacturing techniques. In: Computer aided engineering design and manufacturing: a fourth industrial revolution perspective. Springer, Cham, pp 459–480
43. Khan WA et al (2025) Standards for CAD/CAE/CAM. In: Computer aided engineering design and manufacturing: a fourth industrial revolution perspective. Springer, Cham, pp 369–395
44. Khan WA et al (2025) Mathematics for CAD/CAE/CAM. In: Computer aided engineering design and manufacturing: a fourth industrial revolution perspective. Springer, Cham, pp 357–368
45. Khan WA et al (2025) Use of databases for CAD/CAE/CAM/CIM. In: Computer aided engineering design and manufacturing: a fourth industrial revolution perspective. Springer Cham, pp 397–407
46. Khan WA et al (2025) Communication protocols—legacy and Industry 4.0. In: Computer aided engineering design and manufacturing: a fourth industrial revolution perspective. Springer, Cham, pp 485–504
47. Khan WA et al (2025) Nexus of CIM, Industry 4.0, Industry 5.0, and quantum computing. In: Computer aided engineering design and manufacturing: a fourth industrial revolution perspective. Springer, Cham, pp 551–557
48. Khan WA et al (2025) Quantum computing. In: Computer aided engineering design and manufacturing: a fourth industrial revolution perspective. Springer, Cham, pp 537–549

Part II
Core Technologies and Innovations

Chapter 3
Core Technologies Enabling Industry 5.0

3.1 Objectives

- To specify the technologies that are considered to be the major ones that will allow creating Industry 5.0, such as collaborative robots, AI, IoT, additive manufacturing, and blockchain.
- To discuss the main characteristics of these technologies, their application, and advantages, especially when it comes to the better collaboration between humans and machines, sustainability, and efficiency.
- To evaluate the contribution of AI and machine learning to enhance human decision-making as opposed to replacing human labor.
- To examine how IoT and cyber-physical systems are important in developing real-time and adaptable manufacturing environments.
- To see how blockchain technology will affect transparency, trust and traceability in the supply chain.

3.2 Main Technologies that Support Industry 5.0

The fact is that industry 5.0 is defined by the merging of innovative technologies that seek to improve human imagination, work with machines, and promote sustainability, efficiency, and innovation. Such fundamental technologies allow developing intelligent, flexible, and ethical production systems [1].

M. Hammad et al., *Industry 5.0: The Human-Centric Future of Industry*, Springer Series in Advanced Manufacturing, https://doi.org/10.1007/978-3-032-12622-1_3

3.2.1 *Collaborative Robots (Cobots)*

Collaborative Robots (Cobots) are one of the most well-known technologies that are moving this change. Cobots are made to collaborate with human beings, to increase productivity and safety, flexibility and efficiency in different industrial processes. It is time to discuss the main features of cobots and how they can be used in Industry 5.0 [2].

3.2.1.1 Design

Cobots are collaborative robots, which are designed to work in a close vicinity with human employees. Cobots, in contrast to the traditional industrial robots that are usually sequestered inside a cage or a restricted zone, are supposed to be an integral component of a human team, so they can collaborate easily.

- **High-end Sensors and Force Control Systems**: Cobots have advanced sensors that enable them to sense and react to the surroundings in real-time. These sensors help the robot to know how close it is to people and objects and maneuver its motions to prevent collisions and guarantee safety.
- **Force-Limiting Capabilities**: Cobots have force-limiting abilities in that they are created with a force control system that enables them to exert minimal force when handling human workers. In the case of a human contacting the cobot, the robot will automatically slow down or stop in order to avoid an injury.
- **Adaptive Algorithms**: Cobots have machine learning algorithms that cause them to learn and be able to change within their environment as they go on. These algorithms make cobots more aware of what they have to do, decide in real-time, and optimize their operations, so human–robot collaboration can be effective in different environments.
- **Collision Detection**: Cobots possess a collision detection mechanism that can immediately halt the operations of the robot in case it senses any unwanted contact with a human employee. This aspect enables them to work in dynamic and human-friendly places without any form of protective barriers or fencing [3].

3.2.1.2 Use Cases

The cobots are very flexible and can be implemented in numerous fields and uses. Some of the most common use cases are given below:

- **Assembly Lines**

Role: Cobots help in assembly of products by carrying light weight components, undertaking repetitive activities and supporting high precision activities [4].

Advantages: Cobots improve the quality and uniformity of assembly processes, and the physical load on the employees, to result in better ergonomics and throughput.

Example: In the automobile industry, cobots could be applied in assembling parts of the vehicles, including doors or dashboards, and each component is accurately placed and assembled.

- **Packaging and Logistics**

Use: Cobots are commonly applied to package, sort, and move materials through a factory or warehouse floor. Cobots help to lower the physical burden of workers and enhance efficiency since they perform repetitive and physically demanding tasks [5].

Benefits: Cobots are able to handle heavy lifting and sorting jobs which are physically demanding to human workers and the employees are able to concentrate on more complicated jobs involving creativity or decision making [6–22].

Example: Cobots may help in sorting and packing of products in containers or shipping boxes in a warehouse environment. They are able to carry goods to other places in the warehouse, and this makes the work simple, and the processing time is faster.

- **Healthcare**

Application: Cobots can be applied in many fields of healthcare, including surgical precision and rehabilitation of patients [23].

Advantages: Cobots will be able to help surgeons conduct delicate operations enhancing accuracy and precision of the movements. During rehabilitation, cobots can instruct the patients on how to perform exercises during physical therapy, maintaining the correct posture and minimizing the chances of additional trauma.

Case study: Cobots are applied in surgical environments where they are used to help the surgeons in carrying out the minimally invasive procedures by holding the surgical tools with accuracy and help in assisting the surgeons in the activities that need stabilized hands. During rehabilitation, cobots can assist the patients by taking them through rehabilitating exercises, maintaining the right posture and tracking the improvement.

Figure 3.1 shows a high-tech factory setting with cobots working autonomously, processing items, and interacting with digital systems, emphasizing the role of automation and cobots in modern industrial manufacturing.

3.2.1.3 Benefits

- **Increased Productivity**: Cobots allow the process of manufacturing to be faster and more accurate and minimise the physical and mental burden of human labour. Such a collaborative relationship improves productivity and efficiency.

  Examples: Cobots are able to perform high-speed repetitive tasks on assembly lines so that human workers can be free to perform tasks that are more complex or need creativity. This makes the speed of production improve and at the same time, quality is maintained.

Fig. 3.1 Cobots working in the modern manufacturing environment

- **Safety**: Safety is one of the priorities of cobots. Their ability to limit force and detect collision diminishes chances of accidents or injury. Consequently, cobots can be used in safer working environments than conventional industrial robots that are usually isolated off humans because of safety reasons.

  Example: Cobots can perform repetitive tasks and even dangerous actions, like when working with hot elements, dangerous blades, or other sharp instruments, that can endanger human workers, in such industries as food processing. Their perception around the presence of humans and the capability to react to the same means that workers will not be injured by the actions of the robot [24].
- **Flexibility**

  Cobots are very flexible and can be quickly reprogrammed to do various tasks and hence they are suitable in dynamic production environments with need of flexibility.

  Example: Cobots are easy to reprogram, so in the production environment where products change regularly, it is possible to use it to perform various tasks, e.g., assembly, packaging, or quality checks. This reconfigurability enables firms to rapidly adjust to new product lines without incurring a lot of retooling expenses and new automation systems.
- **Cost-Effectiveness**

  In contrast to industrial robots that are used in traditional industries, and which may need a lot of infrastructures and isolated working areas, cobots are to be used in direct contact with human workers. This eliminates the necessity of huge automation systems or retooling of the whole production lines.

  Example: Cobots can be applied to the factory of small and medium-sized manufacturers without expensive upgrades and adjustments to the existing factory design. It is also possible to retrofit cobots into existing workstations, which means that they are a relatively low-cost solution to the companies that require improving their productivity but do not want to invest as much money as it would take to implement large-scale automation.

3.2.1.4 The Future of Cobots

The use of cobots is going to increase as the Industry 5.0 develops. They are getting integrated into the human–machine cooperative system in which employees are devoted to creative and strategic activities, and cobots perform routine and physically challenging duties. Artificial intelligence and machine learning integration will also increase the capacity of cobots to be more independent, be able to learn based on human behavior, and evolve in the process.

Future: As the cobots develop, they will be more user-friendly and able to perform a broader variety of tasks, and eventually more industries will be able to take advantage of them. A closer partnership between people and computers will result in a more productive process and greater innovation in multiple industries, such as automotive, healthcare, logistics, and consumer electronics [25].

3.2.2 Artificial Intelligence and Machine Learning:

Artificial Intelligence (AI) and Machine Learning (ML) are changing the industry with smart systems that can read, learn about new circumstances, and offer forecasts in terms of patterns. Nevertheless, the main philosophy of Industry 5.0 is based on the idea of the co-making decision instead of substituting human employees with AI frameworks. It is all about enhancing human judgments, enabling people to use the potential of AI to become more creative, efficient, and able to solve problems. AI and ML are smart assistants of human workers, who help them make better and data-driven decisions, but they do not replace the role of human judgment, ethics, and creativity [26].

3.2.2.1 Contribution in Decision-Making:

AI and ML influence the decision creation in numerous industries significantly. Such technologies allow workers to make more accurate, quicker, and better decisions by processing high quantities of information and recognizing the trends that a human being would not notice. Within the context of Industry 5.0, AI is an effective instrument that will help people to expand their abilities, providing decision-makers with new opportunities and insights [27].

- **Predictive Analytics**

Predictive analytics AI and ML technology are commonly utilized to assist organizations to foresee trends and results in the future using previous information. AI systems can predict future challenges or opportunities based on the patterns in previous data and assist the businesses in making more informed decisions.

How It Works: The predictive analytics systems use the past and present data to make predictions concerning events in the future. These predictions can be used by the companies in planning, risk reduction and capturing opportunities.

Case in point: In production, AI-based systems can determine when a machine is prone to break down based on the sensor data of temperature, vibration, and pressure. This enables the businesses to perform maintenance in advance and thus prevent unexpected downtimes and spend money on emergency services.

Example: AI in retailing can be used to forecast the customer demand using past customer purchasing behavior, seasonal patterns, and other external influences like market trend or weather patterns. This enables the companies to change the level of inventory and manage the stocks.

- **Data-Driven Insights**

  Among the most obvious benefits of AI and ML, one can distinguish the capacity to analyze large quantities of data and extract conclusions that can be used by humans to make decisions. The AI systems are useful in detecting inefficiencies, opportunities, and correlations in big data that would require too much time to be discovered by human decision-makers.

  How It Works: AI solutions are capable of examining data that is received by various sources, including production systems, market trends, customer feedback, and social media. The system will then establish patterns, trends and anomalies that can be used to take actions.

  Example: In supply chain management, AI can study historical data on logistics, and determine where the delays could happen, like the possible disturbances in transportation or the lack of stock. This allows the businesses to make proactive adjustments to their operations thereby lowering their costs of doing business and shortening the delivery times. An example: product design can also be assisted through the use of AI whereby customer feedback, reviews and preferences are analyzed. This data can be processed by the AI and propose some design changes or features that can be loved by the customers the most to make the product more in line with the market demand.

- **Collaborative Decision-Making**

  In Industry 5.0, AI will not be left to make decisions on its own, but it is expected to cooperate with human decision-makers. The role of AI is to suggest, recommend and advise people on decisions that can be made by humans, and it is more human-like to make decision-making.

  How It Works: AI systems are programmed to offer suggestions or information depending on the analysis of data, but the decision remains that of the human being. This method enables companies to integrate the advantages of both data-processing power of AI and human wisdom, inventiveness, and moral values.

  Example: In the healthcare field, AI may study the data on patients and suggest possible treatment solutions to them with reference to past medical records and

Fig. 3.2 Visualization depicting AI, machine learning and human cognitive process

studies. Nevertheless, the ultimate decision concerning the treatment process is taken by doctors, and the human judgment (regarding the individual situation and preferences of the patient) is not lost.

Example: In customer service, AI-based chatbots may give recommendations on the response in accordance with past interactions and customer information. Whereas chatbot will process simple questions, human agents will be able to make more complicated choices, which will guarantee a more individual and empathetic experience to customers.

Figure 3.2 shows the synergy between AI and human decision-making, highlighting the balance of machine intelligence with human creativity and judgment.

3.2.2.2 Advantages of AI and Machine Learning in Decisions Making

The adoption of AI and ML in decision-making can have a lot of advantages, including accuracy and efficiency of its operations. Using the capabilities of AI, organizations are able to make better and quicker decisions, thus enhancing their performance and competitiveness [28].

- **Increased accuracy**

  The AI and ML systems are very competent in the analysis of large amounts of data with great accuracy. In comparison to humans who are susceptible to large volumes of information, AI is capable of processing and analyzing intricate information with zero errors and thus businesses can make highly accurate decisions.

Example 1: In financial services, AI can use past market data to predict future movement in the stock market with the accuracy that would be hard to attain by human analysts. This will help the investors to make better decisions regarding the purchase or sale of assets.

Example 2: In quality control in manufacturing, the AI systems can detect minute defects in products that can be overlooked by human inspectors, making products of high quality before they reach the customers.

- **Improved Efficiency**

  The AI and ML technologies can be used to automate routine processes that would otherwise consume human time and energy. AI enables human workers to concentrate on more strategic, creative, and value-added tasks by relieving them of repetitive tasks thus increasing the efficiency of their operations.

  Example 1: In production, AI can perform the sorting of parts according to their specification and leave humans to perform tasks that are more creative like creating new products or coming up with better production processes.

  Example 2: In marketing, AI can be used to automate the data, customer segmentation process, and allow businesses to find the target audiences promptly and personalize the marketing campaigns, whereas human marketers concentrate on strategy and creative development.

- **Quicker Decision-Making**

  The AI systems allow businesses to make quick decisions because they process data in real-time. This is especially relevant in those industries whose processes require timely responses in order to be competitive and adapt to the market changes.

  Example 1: In retail, AI can interpret the data on consumer behavior on a real-time basis and alter the pricing or promotions in real-time to maximize the sales and profits and enable businesses to react to the market trends much faster than the human-driven process would have been possible.

  Example 2: The use of AI in emergency response management: AI can consume real-time data provided by multiple sources (e.g. weather forecasts, traffic reports, social media) to inform authorities to take immediate decisions, i.e. to deploy emergency services or to issue evacuation orders.

AI and ML are important in Industry 5.0 to enhance human decision-making and not to replace them. These technologies assist in streamlining business operations, enhancing quality, and maximizing responsiveness because of the insight on the future they present, data-driven suggestions, and real-time analysis. The secret of success in Industry 5.0 is to establish a symbiosis of human and AI in which the first performs the main part of the work with large amounts of data and makes predictions, and the second one makes the end decision based on his/her judgment, creativity, and moral integrity. The strategy results in a more effective, accurate, and innovative decision-making process that will result in long-term growth and sustainability.

3.2.3 IoT and Cyber-Physical Systems: Real-Time Data Congregation and Adaptive Production

In Industry 5.0, intelligent environments are made up of the integration of advanced technologies. The Internet of Things (IoT) and Cyber-Physical Systems (CPS) are some of the most significant ones. These systems allow a smooth flow of communication among physical machines, sensors and digital networks so that the real-time data can be used to control adaptive and efficient manufacturing processes. Factories and industries can become more independent, smarter and more sustainable through the use of the technologies [29].

3.2.3.1 IoT in 5.0 Industry

The Internet of Things (IoT) is the interconnection of many physical objects that include machines, sensors, and other equipment through the internet. The networked ecosystem allows collecting, communicating, and analyzing data in real-time, which is essential to decision-making and process optimization.

The Functioning of IoT in Industry 5.0

- **Interconnectivity**: IoT enables real-time interconnection of machines and sensors and enables them to exchange information between different systems. This interconnection is the foundation of smart factories in which the machines not only get monitored but also interact with each other to regulate the production processes on their own.
- **Data Collection and Sharing**: Devices that have the capability to be connected to IoT constantly gather information that may comprise of parameters like machine efficiency, environmental factors, speed of production, and inventory levels. This information is then distributed in networks and managers can also follow the operations remotely and make decisions based on current information.
- **Real-Time Monitoring**: IoT enables companies to monitor their activities in real-time by connecting the devices and machines to a central network. This ability gives an instant view of different elements of the manufacturing process, such as the production process itself to supply chain management.

Use Cases of IoT in Industry 5.0

- **Smart Factories**: Smart factories employ the use of machines and devices that communicate independently to correct processes, improve energy consumption and to make sure that production is in full swing. As an example, an IoT-enabled machine may automatically slow down or stop working in case of a problem and avoid system crashes, eliminating the need to use human control.
- **Inventory and Supply Chain Management**: The IoT devices monitor products at all the stages of the supply chain, including the raw material acquisition and the final delivery. The sensors can be used to track the quantity of items, anticipate

the time of shortage and streamline the logistics to minimize time wastage and have the item ready when it is required.

- **Predictive Maintenance**: IoT sensors installed into machines gather information based on performance indicators such as temperature, vibration and pressure. This information enables the businesses to anticipate possible equipment breakdowns, perform preventive maintenance, and eliminate unexpected downtime to enhance the overall efficiency of operations [30].

Example: IoT-connected systems in the automotive manufacturing industry can monitor the condition of the production lines and keep a check on various factors such as availability of parts, machine performance, etc. Such connection also allows that in case a machine starts to malfunction, or a supply problem occurs, the corrective measures can be started right away, and the production process will not be disrupted much.

3.2.3.2 Cyber-Physical Systems

Cyber-Physical Systems (CPS) are the systems in which computer-based algorithms are closely coupled with the physical processes and utilize real-time information to regulate and monitor these processes. CPS also allow more responsive systems, particularly in a manufacturing setting, in which the conditions of operation may vary quickly [31].

The way CPS operates in the Industry 5.0:

Physical and Digital Convergence: CPS are intended to merge physical processes with computer-based algorithms, and those allow real-time control and dynamic adaptation. Integration is very important in the development of responsive manufacturing systems that are able to automatically adapt to changes in the production environment.

Real-Time Adjustm ent and Monitoring: CPS systems allow adjusting the behavior of physical systems (e.g., machines, assembly lines) according to the real-time data of IoT sensors. Examples are, in case a sensor is activated by a bottleneck or a delay, the CPS can automatically speed up a conveyor belt to optimize the workflow.

Autonomous Decision-Making: CPS can make real-time decisions using data that they are fed with and hence can adjust production systems without any human input. These systems may be programmed to take certain priorities, e.g. to minimize energy consumption, waste or throughput.

- **Use Cases of CPS:**

  Adaptive Manufacturing: CPS in adaptive manufacturing systems can be used in optimizing workflows depending on the changing conditions. To illustrate, when the demand of the products increases, CPS will be able to change the speed of production, redistribute resources, and optimize the working process without the supervision of the human factor.

Fig. 3.3 Illustration of IoT and CPS in adaptive manufacturing

Dynamic Process Control: CPS systems are capable of controlling and regulating processes e.g. temperature control in ovens, pressure control in production lines, or material handling according to real time data input and thereby ensuring the best performance and minimizing waste.

Energy Optimization: CPS has the capability of controlling the energy requirements by changing the behavior of machines according to the availability of energy, production plan, and requirements. As an example, when there is low demand, CPS will be able to slow down or switch off machines, which saves energy and makes operations cheaper [32].

Example: A CPS combined with IoT devices in smart manufacturing may be used to speed up or slow down a conveyor belt in real time depending on the information provided by an IoT sensor that counts the items passing through the system. In case the products flow slower than it is expected, the CPS may automatically accelerate the conveyor or indicate the necessity to bring some extra resources to ensure efficiency and minimize the possibility of bottlenecks.

Figure 3.3 depicts the IoT and CPS in manufacturing ecosystem, showcasing real-time data integration and smart factory operations.

3.2.3.3 IoT and CPS Benefits

i. **Real-Time Monitoring**: Using IoT and CPS, it is possible to gain constant real-time insights into the production process enabling organizations to track the performance indicators, identify any abnormalities, and resolve them before they become a major problem. This feature enhances efficiency of operations, minimizes downtime and increases decision-making.

Example: IoT sensors and CPS can be used in smart cities to survey the streetlights, water systems, and traffic, reporting real-time data to streamline the work and enhance citizens services.

ii. **Flexibility and adaptability**: Among the greatest benefits of the IoT and CPS technologies, their capability to automatically regulate the systems according to the real-time information may be pointed out. Such flexibility will enable more flexible manufacturing processes that will be responsive to change in external conditions, market demands, or production requirements.

Example: In an automated factory, when a shift in product specification happens or there is a hiccup in the supply chain, CPS can adjust the production schedules and the machine settings to reduce the effects of the change and have a high throughput.

iii. **Operational Efficiency**: IoT and CPS allow companies to recognize the areas of inefficiency, minimize waste, and better allocate resources based on the continuous gathering of information, processing, and analysis of the collected data. The feedback mechanism enables companies to streamline their operations with time.

Example: In industries that have high energy consumption, CPS can regulate the use of energy by adjusting the machinery according to the real-time cost of energy and this will save the unnecessary consumption of energy, and the cost of operation will also decrease.

Combination of IoT and Cyber-Physical Systems (CPS) is a key to implementing the industry 5.0 through the production of intelligent, adaptive and efficient manufacturing systems. Those technologies offer real-time information that is required to achieve continuous optimization, proactive decision-making, and dynamic process changes. As industries experience the shift towards smarter, more responsive and automated processes, IoT and CPS will further lead to innovation, improved efficiency in operations, and the ability to generate sustainable and adaptive manufacturing environment [33].

3.2.4 Additive Manufacturing: Mass Personalization, Rapid Prototyping and Customization

With its ability to provide new opportunities in product design, production efficiency and customization, additive manufacturing, commonly referred to as 3D printing is transforming industries. Additive manufacturing constructs a product in layers based on a digital model unlike the traditional subtractive manufacturing processes where material is removed as a solid block. The approach is not only used to design elaborate, personalized solutions, but also introduces flexibility and efficiency to the manufacturing procedures in Industry 5.0 [34].

3.2.4.1 Customization and Mass Personalization

Among the most thrilling advantages of additive manufacturing is that it allows the creation of highly specific products that address the exact needs of individual consumers. This is changing a number of industries as it enables companies to go beyond mass production and provide more customized products without losing efficiency or even raising costs.

- **Customization**:

  Additive manufacturing enables production of products that meet the particular needs, be it individual consumer preferences or specialist medical needs. Since 3D printing entails the construction of objects based on digital files, the process of customization becomes more affordable and much simpler to manipulate the design in a short period of time.

  Case in point: 3D printing can be used in the medical sector in the manufacture of personalized prosthetics or implants that are made to fit a specific patient depending on his or her anatomy. This makes fittings more accurate and comfortable compared to the traditional ones and enhances patient outcomes and satisfaction [35].

  Example: In fashion, additive manufacturing enables fashion designers to produce customized clothes and accessories that fit an individual preference, either a custom-fit garment or personalized jewelry. Not only does this make the customers happier but also makes the overproduction of goods less harmful to the environment since goods are produced upon order.

- **Mass Personalization**:

  Whereas the conventional manufacturing techniques have been unable to provide mass production and customized products, additive manufacturing removes this constraint. Through the 3D printing technology, businesses are able to mass produce customized products without much cost and time implication.

  Example: Car manufacturers are employing additive manufacturing to make customized car components and accessories. Instead of having to stock up huge quantities of parts, manufacturers can now make bespoke parts on-demand according to specifications of customers and sell personalized vehicles in large volumes.

  Example: In consumer electronics additive manufacturing can be used to produce customer-specific products or components e.g. phone cases, laptop shells or even customizable interior parts at cost-effective prices so that customers can obtain a unique product to their preferences.

3.2.4.2 Rapid Prototyping

Besides the customization process, 3D printing is also a potent method of rapid prototyping, which improves the speed of the product development process. This enables the companies to test, iterate and refine designs at a fast and efficient pace, saving the time and costs that is normally spent by the companies, using traditional methods of prototyping.

- **Working Principle**: Additive manufacturing allows the engineer and the designer to create working prototypes based on the digital files. Such prototypes can be used in real-life scenarios, where the feedback can be retrieved sooner and more iterations can be made without involving costly molds and tools. This quick cycle of feedback makes product development shorter and innovative [36].

  Examples: 3D printing is used by automotive manufacturers who can rapidly put together prototypes of new automobile parts. To illustrate this, a company can design and test a prototype of a new engine component or interior part and change the design on the fly depending on the performance of the part and only invest in mass production when it is satisfied with the design.

  Examples: Rapid prototyping is used in making and testing new devices by electronics companies. Engineers can produce a new smartphone design in a few days and test various materials, shapes and functions before the final product is ready to be mass produced.

This allows the companies to accelerate the time to market of their products by allowing them to iterate the product design faster than they would be able to with conventional approaches. The conventional methods of prototyping e.g. injection molding or CNC machining can be quite costly as large amounts of money are invested in tools and materials. Additive manufacturing avoids all these expenses and thus the entire process of prototyping becomes very cheap. Prototypes are created out of digital models and thus, the accuracy of the designs is increased and there are lesser mistakes that may be made when they are shifted into physical models.

Figure 3.4 shows the additive manufacturing enabling customization, rapid prototyping, and mass personalization in Industry 5.0. It depicts a 3D printer creating custom products, with symbols for rapid prototyping and personalized design in an industrial setting.

3.2.4.3 Benefits of Additive Manufacturing

The use of additive manufacturing in Industry 5.0 is associated with a number of critical advantages that assist enterprises to enhance efficiency, minimize wastage, and develop innovative goods that satisfy the consumer demand.

i. **Mass customization**: Additive manufacturing helps firms to create personalized products in large quantities without making prohibitive costs. The conventional manufacturing process usually fails to accommodate the need to customize production and the need to be cost effective but with the use of 3D printing, companies can provide personalized products to their customers and still produce large quantities.

   Example: In the manufacturing of medical devices, 3D printing can be used to custom make implants and prosthetics in accordance with individual patients

Fig. 3.4 3D view of additive manufacturing in Industry 5.0

and even make mass production of a broad variety of patients without having to spend too much money [37].

ii. **Cost Efficiency**: Among the greatest benefits of additive manufacturing, one of them is its cost-effectiveness which is manifested in minimizing the wastage of materials and increasing the speed of production. In contrast to subtractive manufacturing processes that need to remove a lot of material out of a solid block, 3D printing only utilizes the quantity of material needed to create a part, leaving fewer unwanted materials.

    Illustration: In the aerospace industry, additive manufacturing enables the manufacturers to create lightweight, intricate components that save the expenses on materials and wastes. Moreover, on-demand production of parts allows decreasing the expenses associated with storage and inventory.

    Example: In consumer goods, 3D printing can be used in reducing the cost of mass producing and transporting products by allowing on-demand local production of customized products.

iii. **Flexibility**: The process of manufacturing becomes highly flexible due to the possibility of designing and developing prototypes quickly and modifying the designs. This flexibility encourages quicker innovation and enables businesses to adapt to the changing market needs or consumer needs in a short period of time.

    Case: In consumer electronics, 3D printing enables the companies to revise the design of the products, e.g., smartphone or wearable, according to the customer feedback or new tendencies, giving the companies an opportunity to create new innovative devices that can satisfy consumer needs.

iv. **Innovation**: Additive manufacturing promotes innovation in all types of industries because it allows them to innovate quickly and create personalized designs. The designers and engineers are able to test the new ideas, materials, and forms without expensive tooling and manufacturing adjustments, and in doing so come up with new products and technologies.

Illustration: In fashion, designers are able to work with special materials, textures, and shapes with the help of 3D printing, and the result is the creation of avant-garde clothes that would not be possible to create with the help of classic technologies [38].

3.2.4.4 Future of Additive Manufacturing in Industry 5.0

The additive manufacturing technology is revolutionary and central to Industry 5.0 since it allows customizing, quick prototyping, and mass personalization. Having the potential to produce highly customized products and save time and cost of production, 3D printing enables businesses to satisfy the increasing need of customized products and ensures efficiency and flexibility in the production process.

With the further development of Industry 5.0, additive manufacturing will become an even bigger part of the innovation processes, efficiency of the operations, and customer satisfaction with custom solutions. The 3D printing is set to transform the future of manufacturing in different industries by integrating the strength of personalization with affordability and quick innovation [39].

3.2.5 Blockchain: Supply Chain Enhancement Through Trust, Traceability and Transparency

One of the pillars in the Industry 5.0 is the blockchain technology which offers better trust, traceability, and transparency in the global supply chain. Blockchain allows sharing and validation of data between multiple parties with the guarantee of data integrity and prevention of fraud using a decentralized and distributed ledger. This not only makes the operations smooth but also brings about teamwork, responsibility and trust in the way business is conducted [40].

3.2.5.1 Role in the Supply Chains

- **Trust**: Among the most important things that blockchain can bring to supply chains, it is possible to note its role in establishing trust between all parties involved. Blockchain will document every transaction in an open ledger that cannot be altered or tampered with after being verified. This implies that the information provided by businesses, suppliers and the consumers can be relied on because it is accurate, verified and cannot be changed.

How It Works: All transactions on the block chain are confirmed by all parties on the network and hence no one party can manipulate data without agreement with the rest. This has a decentralized structure so that the integrity of data is upheld in the supply chain [41].

Case: With the blockchain technology, the food industry can monitor the movement of a product through farm to table. This makes sure that the food that consumers buy is as it is described so as to eliminate fraud, like the food being labelled to be organic when it is not or having ingredients that are not safe.

- **Traceability**: The traceability capability of blockchain enables the stakeholders to trace the products in all the steps of the supply chain, the raw material supply chain to the final delivery. This full visibility is especially necessary in the industries where products safety, authenticity, and compliance are the most important factors.

  How It Works: Using blockchain, all the processes of supply chain are documented, and each transaction is tied to the identity of a product. Be it the origin of raw materials, the components in the manufacturing processes, or authenticity of a finished product, blockchain enables a hitherto impossible level of traceability.

  Examples: In pharmaceutical industry, blockchain can be used to monitor the flow of drugs between the manufacturers, wholesalers and pharmacies, and therefore guarantee that drugs have not been altered or counterfeited. It is important in avoiding sale of counterfeit drugs and safety of patients. In another example, while considering the luxury goods, blockchain may be used to verify the origin of expensive luxury goods such as diamonds or luxury watches, and it allows customers to be confident that they are buying a real and high-quality product and not a fake.

- **Transparency:** The decentralized aspect of blockchain is that this information is available to all the parties involved in the process, i.e., suppliers, distributors, and consumers in real-time. Such a degree of transparency is priceless in terms of trustworthiness, minimization of conflicts, and development of business ethics.

  How It Works: All transactions on the blockchain are publicly documented (although privacy and security are ensured by encryption), which means that all the stakeholders have equal access to the history of transactions. This would enable them to confirm the product information, the sources of products used as well as any modification done during the supply chain process.

  Examples: In fair trade blockchain can be used to ensure fair wages are paid to the workers in the production process and that the products are ethically sourced. Blockchain enables ethical consumerism and helps build brand integrity by giving clear information on labor practices and sourcing. Another example is that consumers are becoming more concerned about the environmental impact of the products they buy. Blockchain offers transparency in that businesses can demonstrate their sustainability assertions, including the carbon footprint of a commodity, materials, or environmental practices of the manufacturers.

Figure 3.5 shows the supply chain components like trucks, factories, and product packaging, emphasizing trust, traceability, and transparency in supply chains.

Fig. 3.5 Blockchain enhances supply chain transparency, traceability, and trust in Industry 5.0

3.2.5.2 Benefits of Blockchain

The transparency, trust and traceability that the blockchain offers come with a lot of benefits throughout supply chains. These advantages do not only enhance efficiency in operations but also instill consumer confidence and stimulate ethical practices.

i. **Better Supply chain visibility**: It is possible to track products, materials, and components in the supply chain in real-time through Blockchain. Such increased visibility gives a better understanding of possible bottlenecks, risks, or interruptions that may affect the operations.

   Retail: Blockchain will be able to give visibility to the inventory levels and allow businesses to monitor products within warehouses and distribution centers in real time. This enhances control of inventory and minimizes cases of out-of-standard and customer satisfaction.

   Example: In car manufacturing, blockchain can be used to trace the flow of car parts in the global supply chains, so that they are available when required, and there is a minimum of delays and surplus inventory, and costs are reduced [42].

ii. **Minimized Fraud and mistakes**

   Blockchain can lower the risk of fraudulent practices, as it guarantees the integrity of data with the help of its immutable ledger, e.g. counterfeiting, mislabeling, or tampering with information about the product. This is especially relevant to those industries where authenticity of products matters a lot.

   Examples: In fashion, blockchain will make it possible to guarantee that luxury brands can demonstrate the authenticity of their goods, which is why they cannot be counterfeited, and consumers will receive real, high-quality products.

This openness contributes to the preservation of brand value and the increase in customer loyalty. Another example in food safety, blockchain will allow preventing fraud by confirming that organic or sustainably produced food is not labeled as conventional. This minimizes the chances of consumer boycotts and loss of reputation of businesses.

iii. **Growth of Consumer Confidence**

The consumers have become more sensitive about the source, health and authenticity of the goods they are buying. Transparency on blockchain gives the consumers access to quality information about products, which increases their confidence and trust in purchased goods.

Case: Food and beverages: Consumers will be able to scan the labels of products with apps based on blockchain to check the details of the product including origin, ingredients and production process. This creates confidence and promotes consumer loyalty, especially in the rising demand for organic and ethically produced products.

Example: In the high-end goods market, it is possible to verify the origin of luxury items such as watches, handbags, or jewelry, and the consumer is sure that they are buying a genuine product and contributing to the sustainable brands.

The evolution of Industry 5.0 can be predicted to use blockchain to facilitate the development of transparent, trusted and traceable supply chains that will promote innovation, efficiency and sustainability to the overall benefit of businesses, and consumers.

3.3 Technology Readiness Level and Integration Issues

The industry 5.0 is rapidly changing, and it is necessary to comprehend the maturity of the technologies adopted to achieve successful integration. Technology Readiness Level (TRL) is one of the most important parameters that are used to gauge the maturity of a technology. This scale, which goes to TRL 1 (the basic principles observed) to TRL 9 (actual system proven through successful mission operations), allows to estimate the possibility of new technologies to be implemented into industrial systems [43].

- **Technology Readiness Level (TRLs)**: With the transition of industries to the Industry 5.0, the technologies such as Artificial Intelligence (AI), Collaborative Robots (Cobots), Blockchain, and Cyber-Physical Systems (CPS) are being deployed to manufacturing facilities. The maturity of these technologies and the difficulties that are involved in integrating them is a major factor in determining the success of adoption and implementation of these technologies.

Knowledge of the TRLs of major Industry 5.0 technologies assists the organizations in determining whether the technologies are at a stage that they can be deployed at a full scale or that they need additional development before they can be utilized in industrial systems.

TRL 1: The basic principles are noted and described. On this level, research is at its initial stages, and the possible use of the technology is not proved.

TRL 2: Concept and application of the technology is formulated. Initial tests have been done to illustrate the possibility of technology.

TRL 3: Demonstration of proof of concept. The preliminary study and testing are effective to demonstrate the possibility of technology.

TRL 4: Technology is lab proven. The technology is applied in controlled settings; however, it is not established in a real-life industrial environment.

TRL 5: Technological demonstration in pertinent environment. Technology has been tried in a realistic scenario which resembles the industrial environment.

TRL 6: Technology that is proven in pertinent environment. Technology is completely shown in an artificial real-life scenario.

TRL 7: Prototype of the system is tested in the working environment. A real-world industrial testing of a working prototype of the system is done.

TRL 8: Completed the actual system and ran. It is an industrial system that is fully integrated and works with the requirements of performance.

TRL 9: Real system demonstrated by successful operations of missions. The technology is ready, and it has been implemented successfully in practice.

In the case of Industry 5.0 technologies, it is paramount to know their TRL levels in order to be able to tell whether they are ready to be implemented in a mass way or they still require improvement. The technologies at TRL 6–9 are usually those which are set to be implemented at full scale and the ones at lower TRLs need to be developed further before they can be incorporated into the industrial systems.

3.3.1 Challenges in Integration

Although Industry 5.0 technologies have potential, they are facing a number of challenges regarding their integration into the current industrial situation. To overcome these challenges, it is necessary to plan, invest and be ready to learn and develop constantly.

i. **Complextity of Integration**: The adoption of new technologies such as AI, cobots, and blockchain in the current manufacturing system is a complicated procedure that necessitates a lot of changes in the infrastructure, process, and organizational culture.

   Infrastructure Adjustments: Infrastructure already installed might not be able to cope with the escalated needs of new technologies. As an example, applying AI and machine learning, is associated with strong computing resources, storage, and network, and not all the traditional systems may have it.

Processes: The incorporation of certain technologies, e.g., cobots (those that cooperate with human employees) may imply a need to alter workflows and to modify factory design to ensure that human and machine interaction is safe. Also processes based on legacy systems may have to be modified to interact with new more automated systems and thus there may be disruptions in the transition period.

Examples: In a typical production facility, implementation of collaborative robots (cobots) might necessitate restructuring of production processes, e.g. modification of assembly lines to fit robots that will help human beings to undertake monotonous tasks.

Solution: Applying AI in the industrial environment not only implies installing software and hardware but also re-engineering the current workflows and adequate integration of AI systems into the business-decision making process.

ii. **Interoperability**: To be effective, the Industry 5.0 technologies must be able to work with the legacy systems. Interoperability of new technologies (such as IoT, blockchain, AI, and cobots) with the older industrial systems is one of the greatest tasks of the contemporary manufacturing.

Interoperability Problems: The old systems that can be decades old are usually not made to operate in the modern systems that operate on open-source software, cloud computing, and real time data sharing. This causes problems when new systems are to be incorporated in the old production lines.

Example: A lot of factories use legacy software or manual systems to manage the inventory. The incorporation of blockchain and IoT to enhance the traceability of the supply chain may entail a lot of modifications to these outdated systems, which is both time-consuming and costly.

Solution: One way to address the problem of interoperability is the implementation of a middleware solution or IoT platform that helps to connect old systems with new ones and provide easy data flow between systems.

iii. **Workforce Readiness**: The shift to Industry 5.0 demands a human resource, which is competent in the technical part of new technologies and the human-centered approach which Industry 5.0 promotes. Although automation and AI increase the efficiency of operations, human creativity, problem-solving, and decision-making are fundamental attributes of the workforce.

Training and Upskilling: Industry 5.0 is based on the interrelation of human skills and high-tech technologies. It implies that the employees should be constantly trained to work in new systems, and the constant dedication to the upskilling of the workers should be in place to keep up with the development of technologies.

Human-Centered Approach: The idea of human-centered approach is also an attribute of Industry 5.0. Employees should not only have technical skills but also the skill to coordinate and work with autonomous systems such as AI, cobots, and IoT technologies.

Example: In the case of AI and machine learning being applied in manufacturing, employees would have to be trained in how to communicate with these systems and rely on AI-assisted conclusions. The workforce has to be changed to cooperate in the environment, in which human beings collaborate with robots and other autonomous systems.

Solution: The constant training process and upskilling programs are the key to keeping the workforce able to cooperate successfully with the changes in technology. Furthermore, the creation of the culture of life-long learning will allow businesses to develop a workforce that can easily enter the era of the Industry 5.0.

Businesses have to succeed in formulating a clear plan for integrating new technologies, making infrastructure investments and employee training, and systems compatibility between legacy and new systems. In such a way, the companies will be able to unlock the potential of the Industry 5.0 technologies and generate more agile, efficient, and innovative operations [44].

3.4 Chapter Learning Outcomes

After reading this chapter, the readers will have an opportunity to:

- Learn the main technologies which are in the driving force of Industry 5.0 and their particular contribution to the improvement of human creativity and collaboration.
- Determine primary use cases of collaborative robots, AI, IoT, additive manufacturing, and blockchain in the contemporary industries and its advantages.
- Examine the consequences of AI and machine learning to the decision-making process and the way in which it enhances human abilities.
- Evaluate the prospects of IoTs and cyber-physical systems in real-time data aggregation and manufacturing processes that are adaptive.
- Explain how blockchain can improve transparency, traceability and trust in the supply chains.
- Understand the difficulty of incorporating the Industry 5.0 technologies in the current systems and the significance of being prepared in the workforce.

3.5 Review Questions

1. How do collaborative robots (cobots) transform the workplace in Industry 5.0 compared to traditional industrial robots?
2. (a) In what ways do artificial intelligence (AI) and machine learning (ML) contribute to enhancing human decision-making in Industry 5.0?

   (b) Provide specific examples of predictive analytics, data-driven insights, and collaborative decision-making in industrial settings.
3. (a) Critically evaluate the role of the Internet of Things (IoT) and Cyber-Physical Systems (CPS) in adaptive manufacturing.

(b) How do these technologies contribute to real-time data integration, resource optimization, and autonomous decision-making in modern production systems?

4. (a) Discuss the potential of additive manufacturing (3D printing) in Industry 5.0. How does it enable customization at scale and reduce material waste?

   (b) What are the challenges associated with integrating additive manufacturing into traditional manufacturing workflows?

5. (a) Explain how blockchain technology enhances transparency, traceability, and trust in supply chains within the context of Industry 5.0.

   (b) Provide an analysis of its applications in industries such as pharmaceuticals, food safety, and luxury goods.

6. (a) How do Technology Readiness Levels (TRLs) help in assessing the feasibility of integrating these technologies into industrial systems, and

   (b) What challenges do companies face in achieving full integration?

7. (a) How does the transition to Industry 5.0 address the limitations and challenges posed by Industry 4.0, particularly in terms of human labor displacement and over-reliance on automation?

   (b) Discuss the opportunities for reskilling workers and creating new roles that focus on creativity and innovation.

8. How can businesses effectively integrate cobots, AI, and other advanced technologies to foster collaboration between human workers and machines in production environments?

9. (a) Examine the operational benefits of IoT and CPS in reducing downtime and improving efficiency in industrial systems.

   (b) How do these technologies enable proactive maintenance and dynamic adaptation to change production requirements?

10. (a) Discuss the environmental impact of additive manufacturing in Industry 5.0. How does 3D printing contribute to sustainability?

    (b) What are the potential ecological challenges associated with widespread adoption of this technology?

Appendix: Solution Manual

1. **How do collaborative robots (cobots) transform the workplace in Industry 5.0 compared to traditional industrial robots?**

   Solution: Collaborative robots (cobots) differ from traditional industrial robots by their ability to work safely alongside humans in shared spaces, eliminating the need for safety barriers. Cobots are equipped with advanced sensors and force control systems to detect human presence and avoid accidents. Use cases include assisting in assembly lines, packaging, and healthcare, where they enhance precision, reduce worker fatigue, and increase productivity. Their benefits also include flexibility, as they can quickly adapt to new tasks and are cost-effective compared to large-scale automation systems.

2. **(a) In what ways do artificial intelligence (AI) and machine learning (ML) contribute to enhancing human decision-making in Industry 5.0?**

   Solution: AI and ML augment human decision-making by providing predictive analytics that forecasts future trends, such as demand or maintenance needs. These systems analyze large datasets to offer actionable insights for optimizing operations.

   (b) Provide specific examples of predictive analytics, data-driven insights, and collaborative decision-making in industrial settings.

   Solution: For example, AI-driven systems can suggest supply chain improvements based on real-time data, or recommend design changes based on consumer feedback. AI works alongside humans to provide recommendations, but the final decision remains with the human operator, ensuring creativity and judgment are not compromised.

3. **(a) Critically evaluate the role of the Internet of Things (IoT) and Cyber-Physical Systems (CPS) in adaptive manufacturing.**

   Solution: IoT and CPS enable real-time monitoring of production processes by connecting machines, sensors, and devices. These systems provide data that can be analyzed to make proactive decisions, such as adjusting production speeds or predicting maintenance needs.

   (b) How do these technologies contribute to real-time data integration, resource optimization, and autonomous decision-making in modern production systems?

   Solution: IoT facilitates smart factories where devices communicate autonomously, optimizing resource usage. CPS enables these systems to respond dynamically to changes, reducing human error and ensuring operational efficiency.

4. **(a) Discuss the potential of additive manufacturing (3D printing) in Industry 5.0. How does it enable customization at scale and reduce material waste?**

   Solution: Additive manufacturing allows for mass customization without increasing production costs. It enables companies to create personalized products at scale, whether in fashion, healthcare, or automotive sectors.

   (b) What are the challenges associated with integrating additive manufacturing into traditional manufacturing workflows?

   Solution: However, challenges include the need for specialized knowledge in 3D design and the potential for slower production speeds compared to traditional methods. Integration into traditional workflows requires significant adjustments, such as redesigning supply chains to accommodate 3D printing capabilities.

5. **(a) Explain how blockchain technology enhances transparency, traceability, and trust in supply chains within the context of Industry 5.0.**

Solution: Blockchain enhances supply chain transparency by providing an immutable, decentralized ledger that tracks products through every stage of the supply chain.

(b) Provide an analysis of its applications in industries such as pharmaceuticals, food safety, and luxury goods.

Solution: This is particularly useful in industries where product authenticity and ethical sourcing are paramount, such as pharmaceuticals and luxury goods. Blockchain ensures that every transaction is recorded, reducing fraud and errors and enhancing consumer trust.

6. **(a) How do Technology Readiness Levels (TRLs) help in assessing the feasibility of integrating these technologies into industrial systems?**

   Solution: TRLs measure the maturity of a technology from basic principles (TRL 1) to fully operational systems (TRL 9). Understanding the TRL of technologies like AI, cobots, and blockchain helps companies assess the feasibility of integration into industrial systems.

   (b) What challenges do companies face in achieving full integration?

   Solution: The main challenges include achieving interoperability between new and legacy systems and ensuring workforce readiness to adopt these technologies.

7. **(a) How does the transition to Industry 5.0 address the limitations and challenges posed by Industry 4.0, particularly in terms of human labor displacement and over-reliance on automation?**

   Solution: Industry 5.0 addresses the job displacement caused by automation in Industry 4.0 by emphasizing human creativity and collaboration with machines.

   (b) Discuss the opportunities for reskilling workers and creating new roles that focus on creativity and innovation.

   Solution: New roles will emerge that focus on innovation, problem-solving, and creative tasks. Businesses must invest in reskilling workers and fostering a culture of continuous learning to ensure a smooth transition to a more human-centered industrial model.

8. **How can businesses effectively integrate cobots, AI, and other advanced technologies to foster collaboration between human workers and machines in production environments?**

   Solution: To foster collaboration, businesses should integrate cobots and AI in ways that complement human skills. Cobots can take over repetitive or hazardous tasks, freeing human workers to focus on higher-value activities like problem-solving and innovation. By promoting a collaborative environment, businesses can enhance productivity, safety, and worker satisfaction while ensuring that technology serves as an aid rather than a replacement for human labor.

9. **(a) Examine the operational benefits of IoT and CPS in reducing downtime and improving efficiency in industrial systems**.

   Solution: IoT and CPS enable proactive maintenance by continuously collecting data and identifying potential issues before they lead to downtime.

(b) How do these technologies enable proactive maintenance and dynamic adaptation to change production requirements?

Solution: These technologies allow for real-time adjustments, such as optimizing production speed or reallocating resources based on sensor data. This results in improved operational efficiency and reduced production delays.

10. **(a) Discuss the environmental impact of additive manufacturing in Industry 5.0. How does 3D printing contribute to sustainability?**

    Solution: Additive manufacturing contributes to sustainability by reducing material waste, as it only uses the amount of material required to build each product. It also enables the recycling of materials and supports the production of eco-friendly products.

    (b) What are the potential ecological challenges associated with widespread adoption of this technology?

    Solution: However, challenges include the energy consumption associated with 3D printing and the environmental impact of certain materials used in the process. Widespread adoption must include efforts to reduce these ecological impacts.

Bibliography

1. Nahavandi S (2019) Industry 5.0—a human-centric solution. Sustainability 11(16):4371
2. Rahman MdM et al (2024) Cobotics: the evolving roles and prospects of next-generation collaborative robots in Industry 5.0. J Robot 1:2918089
3. Hammad M et al (2023) Framework for the implementation of smart manufacturing systems: a case in point. Processes 11(5):1436
4. Matheson E et al (2019) Human–robot collaboration in manufacturing applications: a review. Robotics 8(4):100
5. Keshvarparast Ali et al (2024) Collaborative robots in manufacturing and assembly systems: literature review and future research agenda. J Intell Manuf 35(5):2065–2118
6. Khan WA et al (2025) Industry 5.0. In: Computer aided engineering design and manufacturing: a fourth industrial revolution perspective. Springer, Cham, pp 523–535
7. Khan WA et al (2025) Computer aided manufacturing. In: Computer aided engineering design and manufacturing: a fourth industrial revolution perspective. Springer, Cham, pp 283–297
8. Khan WA et al (2025) Production planning and control. In: Computer aided engineering design and manufacturing: a fourth industrial revolution perspective. Springer, Cham, pp 425–453
9. Khan WA et al (2025) Manufacturing and production. In: Computer aided engineering design and manufacturing: a fourth industrial revolution perspective. Springer, Cham, pp 65–185
10. Hammad M et al (2025) Lightweight and robust key agreement for securing IIoT-driven flexible manufacturing systems. IEEE Internet Things J
11. Khan WA et al (2025) Computer-aided numerical analysis techniques. In: Computer aided engineering design and manufacturing: a fourth industrial revolution perspective. Springer, Cham, pp 305–355
12. Khan WA et al (2025) Engineering design. In: Computer aided engineering design and manufacturing: a fourth industrial revolution perspective. Springer, Cham, pp 11–63
13. Khan WA et al (2025) Pillars of Industry 4.0 and CAD/CAM. In: computer aided engineering design and manufacturing: a fourth industrial revolution perspective. Springer, Cham, pp 509–522

14. Hammad M et al (2023) A provable secure and efficient authentication framework for smart manufacturing industry. IEEE Access 11:67626–67639
15. Khan WA et al (2025) Remanufacturing. In: Computer aided engineering design and manufacturing: a fourth industrial revolution perspective. Springer, Cham, pp 187–211
16. Khan WA et al (2025) Future manufacturing techniques. In: Computer aided engineering design and manufacturing: a fourth industrial revolution perspective. Springer, Cham, pp 459–480
17. Khan WA et al (2025) Standards for CAD/CAE/CAM. In: Computer aided engineering design and manufacturing: a fourth industrial revolution perspective. Springer, Cham, pp 369–395
18. Khan WA et al (2025) Mathematics for CAD/CAE/CAM. In: Computer aided engineering design and manufacturing: a fourth industrial revolution perspective. Springer, Cham, pp 357–368
19. Khan WA et al (2025) Use of databases for CAD/CAE/CAM/CIM. In: Computer aided engineering design and manufacturing: a fourth industrial revolution perspective. Springer, Cham, pp 397–407
20. Khan WA et al (2025) Communication protocols—legacy and Industry 4.0. In: Computer aided engineering design and manufacturing: a fourth industrial revolution perspective. Springer, Cham, pp 485–504
21. Khan WA et al (2025) Nexus of CIM, Industry 4.0, Industry 5.0, and quantum computing. Computer aided engineering design and manufacturing: a fourth industrial revolution perspective. Springer, Cham, pp 551–557
22. Khan WA et al (2025) Quantum computing. In: Computer aided engineering design and manufacturing: a fourth industrial revolution perspective. Springer, Cham, pp 537–549
23. Hornych Y et al (2020) Near-ultrasonic communications for IoT applications using android smartphone. In: 2020 IEEE International conference on electro information technology (EIT). IEEE
24. Wallén J (2008) The history of the industrial robot. Linköping University Electronic Press
25. Bauer A, Wollherr D, Buss M (2008) Human–robot collaboration: a survey. Int J Humanoid Rob 5(01):47–66
26. Ahmed I, Jeon G, Piccialli F (2022) From artificial intelligence to explainable artificial intelligence in industry 4.0: a survey on what, how, and where. IEEE Trans Industr Inf 18(8):5031–5042
27. Chong SY, Tan MK, White JD (2005) Observing the evolution of neural networks learning to play the game of Othello. IEEE Trans Evolution Comput 9(3):240–251
28. Hammad M et al (2025) Introduction to smart manufacturing in the era of Industry 4.0. In: Smart manufacturing blueprint: navigating Industry 4.0 across diverse sectors. Springer, Cham, pp 3–18
29. Lu Y (2017) Industry 4.0: a survey on technologies, applications and open research issues. J Ind Inf Integr 6:1–10
30. Qin J, Liu Y, Grosvenor R (2016) A categorical framework of manufacturing for industry 4.0 and beyond. Procedia CIRP 52:173–178
31. Lee EA (2008) Cyber physical systems: design challenges. In: 2008 11th IEEE international symposium on object and component-oriented real-time distributed computing (ISORC). IEEE
32. Derler P, Lee EA, Vincentelli AS (2011) Modeling cyber–physical systems. Proc IEEE 100(1):13–28
33. Da Xu L, He W, Li S (2014) Internet of things in industries: a survey. IEEE Trans Indus Inform 10(4):2233–2243
34. Fianko SK et al (2025) Mass customisation strategies in additive manufacturing: a systematic review and implementation framework. Processes 13(6):1855
35. You X, Yang Y, Guo P (2020) Fabrication of microscale polymeric wavy and coiling structures via side-electrode-assisted near-field electrospinning: modeling and experiments. J Manuf Sci Eng 142(7):071005
36. Roy M, Wodo O (2020) Data-driven modeling of thermal history in additive manufacturing. Addit Manuf 32:101017

37. Huang S et al (2022) Industry 5.0 and Society 5.0—comparison, complementation and co-evolution. J Manuf Syst 64:424–428
38. Espalin D et al (2014) 3D Printing multifunctionality: structures with electronics. Int J Adv Manuf Technol 72(5):963–978
39. Berman B (2012) 3-D printing: the new industrial revolution. Bus Horiz 55(2):155–162
40. Casino F, Dasaklis TK, Patsakis C (2019) A systematic literature review of blockchain-based applications: current status, classification and open issues. Telematics Inform 36:55–81
41. Toyoda K et al (2017) A novel blockchain-based product ownership management system (POMS) for anti-counterfeits in the post supply chain. IEEE Access 5:17465–17477
42. Saberi S et al (2019) Blockchain technology and its relationships to sustainable supply chain management. Int J Prod Res 57(7):2117–2135
43. Koonce D, Sormaz D (2019) Metrics to gauge the success of a manufacturing ontology. Procedia Manuf 38:1678–1682
44. Sakib N, Wuest T (2018) Challenges and opportunities of condition-based predictive maintenance: a review. Procedia CIRP 78:267–272

Chapter 4
Digital Twins and Smart Environments

4.1 Objectives

- To explain the meaning of the notion of digital twins and outline their architecture and the way they work.
- To discuss such possible uses of digital twins in Industry 5.0 as predictive maintenance, process optimization, and real-time monitoring.
- To examine the main advantages of the usage of digital twins, including the minimization of downtime, the increase of product quality, and customization.
- To evaluate the issues related to the implementation and scaling of digital twins, paying attention to the aspects of ROI and complexity of integration.
- To discuss the place of digital twins in smart surroundings and the way they help businesses to streamline operations, cut costs, and be more sustainable.

4.2 Introduction

Digital Twins are quickly changing the paradigm of industries by producing simulated versions of real objects, systems or processes [1]. Those digital twins allow real-time, dynamic models of the real world, allowing companies to observe, analyze, and optimize their activity more effectively. Digital Twins in Industry 5.0 are important to increase human–machine interaction, predictive maintenance, real-time monitoring, process optimization, and sustainability. This section is going to explore the concept, architecture, uses, advantages, and limitations of having Digital Twins in smart environments [2–4].

M. Hammad et al., *Industry 5.0: The Human-Centric Future of Industry*, Springer Series in Advanced Manufacturing, https://doi.org/10.1007/978-3-032-12622-1_4

4.3 Concept and Architecture

A Digital Twin represents a virtual representation or simulation of a real object, process, or system. It constantly gets real-time information about different sensors, IoT devices, and other sources that represent the current state and performance of its physical analog. With this information, the Digital Twin can simulate and model the behavior of the physical objects and processes and enable the companies to make data-driven decisions that enhance efficiency, productivity, and sustainability [1]. A Digital Twin architecture generally comprises a number of important elements that are synchronized to establish a smooth bridge between the virtual and physical model:

- **Physical Object**: The physical object is the real-life object which is being copied into the digital space. This may be a machine, vehicle or equipment, or even an entire industrial process or building or infrastructure. The physical object is the basis of its Digital Twin development that reflects its properties, behaviors, and performance virtually [5].

  Example: A robot used in the manufacturing industry may possess a Digital Twin that replicates its state of operation such as speed, how it is moving and how efficiently it can perform its duties.
- **Data Collection and Sensors**: Data collection is an activity that involves the process of gathering and accumulating data. Data collection is done before the analysis of the data and the interpretation of the data. Data collection can be done in two ways, primary data collection and secondary data collection. Primary data collection involves collecting data using the direct method of data collection. Primary data collection is done through the use of sensors.

  The physical object has IoT devices and sensors collecting real-time data of parameters like temperature, pressure, speed, vibration, energy consumption, and other performance parameters. Such sensors constantly send this information to Digital Twin, so that the virtual model is always updated with the most recent information [6–9].

  Example: A factory scenario may include sensors that can detect the temperature and wear of some of the vital parts of the machinery and pass this information to the Digital Twin to determine whether it requires any form of maintenance.
- **Simulation and modeling**: The data is then passed into a digital simulation and modeling environment once the data has been gathered. This is a step through which dynamic models are created to predict the behaviors, interactions and performance of the physical object. High-tech solutions, such as AI and machine learning, are often applied at this point to increase the predictive power of the model, so that it can be used to simulate different scenarios, e.g. stress testing or operational changes [10].

  Example: A Digital Twin of a wind turbine may be used to model the behaviour of the turbine in varied wind speeds or weather and thus enable the engineer to optimise the performance and reduce risk.

- **Real-Time Analytics**: Digital Twins make use of real-time analytics to always observe the performance of the physical object or system. Digital Twins allow analyzing the data in real-time, which makes it possible to define inefficiencies, anticipate possible failures, and propose data-driven process optimizations [11].

  Example: The Digital Twin of a power plant turbine can help notice the signs of wear or malfunction by constantly monitoring the level of vibration, temperature, and pressure. This enables the predictive maintenance to be carried out before the failure, thus minimizing the down time and the cost of repair.
- **Integration and Feedback**: The last element of a Digital Twin that is critical is the feedback loop and integration between the physical object and the virtual model. Digital Twin always communicates with the physical system in a closed-loop manner, and the information is continuously shared. This will ensure that any changes or optimizations required will be made in real time and ensure that there is a feedback loop that keeps the system operating at optimum efficiency [12].

  Example: At an automated factory, the Digital Twin of the production line is continuously tuned according to performance data that is provided by the sensors. In case a machine is not running as fast as anticipated, the system can automatically change machine settings or request maintenance to fix the problem.

The key components of digital twin infrastructure are shown in Fig. 4.1. Those components include physical objects, simulation modeling and real-time analytics.

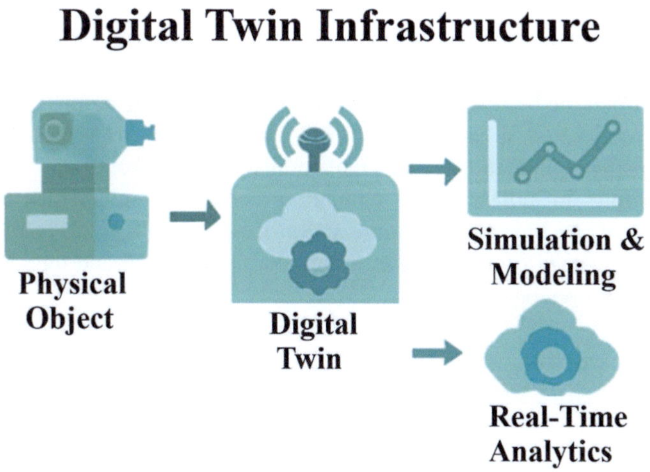

Fig. 4.1 Key components of digital twin infrastructure

4.3.1 Important Elements of Digital Twin

Digital Twins need multiple technological elements to operate in the most efficient way, gathering, processing, and interpreting the data of the physical systems [13].

- **IoT sensors and Devices**: The main activity that is used in the physical world to collect real-time data is the use of IoT sensors. Such sensors are able to record an extensive amount of data such as temperature, humidity, motion, pressure, and vibration. The information gathered with the help of these devices is used as a source of the input of the Digital Twin, which allows it to precisely simulate the performance of its physical copy [14].

  Example: In intelligent cities, IoT sensors that are installed on the street-lights, roads and buildings measure the traffic flow, energy usage and environmental status. Such data is transmitted to Digital Twins of urban infrastructure to streamline management of the city.
- **Cloud Computing**: Digital Twins create tremendous volumes of real-time data that are usually stored, processed, and analyzed using cloud infrastructure. The enormous amount of data requires the computational capacity needed that can be achieved by cloud platforms and they are scalable to large systems. This permits organizations to access their Digital Twins wherever they are and allow real-time analysis of various systems.

  Example: Cloud computing can be used in a global supply chain management system to gather and analyze data in warehouses, transportation vehicles and distribution centres and optimize inventory and minimize delays.
- **Edge Computing**: In others, edge computing can be employed in order to perform the processing of the data nearer the point of origin e.g. on-site at manufacturing plants or industrial sites. This shortens the latency and improves real-time decision-making because the data is processed at the edge before it is transmitted to the cloud where it can be analyzed. Edge computing is especially helpful in time sensitive applications where rapid adjustments are required.

  Example: In industry, edge computing could be implemented to supervise the work of machinery in real-time and process the data at the point where it is generated instead of sending the entire data to the cloud and then analyzing it, which would not be very quick.

4.3.2 Working Principle

Digital Twins work by always getting the information of their real-life counterparts via IoT sensors and other sources of information. This information is fed to update and refresh the virtual model and make sure that the virtual model is always updated to represent the state of the physical system. The loop of the physical object and the digital object allows monitoring the processes in real-time, optimization, predictive maintenance, and testing of scenarios [15–19].

Example: In a smart factory, a Digital Twin of a production line is fed with the data of several machines and sensors to give information about the production rates, energy consumption, and equipment conditions. That data is then used to make real time adjustments to the production process, improving efficiency and reducing waste.

By making sure that Digital Twins constantly reflects the operations and the state of physical systems, businesses will be able to optimize their work, avoid failures, and develop more sustainable and efficient processes.

4.4 Applications for Digital Twins:

The use of Digital Twins in smart environments cuts a broad range of applications across the industries. Digital Twins provide businesses with the opportunity to streamline operations, enhance efficiency, and cut costs since they produce real-time digital versions of physical things, systems, and processes. The most influential uses of Digital Twins are the following:

i. **Predictive Maintenance**

Digital Twins have one of the most effective uses in predictive maintenance. Through constant monitoring of physical assets health and performance, Digital Twins are able to forecast when an asset needs maintenance or when a machine is about to break down. This preventive maintenance strategy assists companies to develop efficient maintenance schedules, prevent untimely shutdowns, and save unwarranted repair expenses [20].

How it Works: Real-time data is captured by sensors installed within a machine or a system and then sent to a Digital Twin. This model performs analysis of data by predictive algorithms to predict the future failures prior to their occurrence. Maintenance activities can then be planned on actual condition rather than on a scheduled basis, thus the cost is saved, and equipment life is extended.

Example: Digital Twins in the automotive industry are utilized to track the state of the important parts of the car such as the engine, brakes, and tires. As an example, a Digital Twin of an engine might be used to study temperature variations, vibration rates, etc. to foresee wear and tear, and raise warnings to operators on the verge of some problems so that they could be fixed before causing costly repairs. In addition, in the aviation industry, Digital Twins of parts of the aircraft- including engines, tires and avionics can monitor their health in real-time. The predictive maintenance functions mean that the components can be swapped or fixed before they break, decreasing unscheduled outage and enhancing safety. When these components are tracked through their entire lifecycle, the operators will be in a position to prevent expensive flight schedule delays.

ii. **Process Optimization**

Digital Twins can be used to optimize the process because through them the business can simulate, test and improve a process prior to deploying it into the real world. Virtual modeling allows companies to test the alterations of their production line, supply chain or service process and determine where to make it better and also make sure that the changes are efficient and effective.

How it Works: It allows businesses to model an entire process in a digital interface so that they can simulate different scenarios and see the possible consequences of their actions, without interfering with operations. This enables companies to experiment, streamline processes and discover bottlenecks without making any changes in the physical environment [21–25].

Example: In the manufacturing industry, Digital Twins of production lines can model the process of the whole workflow to find out inefficiencies, including delays that are unnecessary or waste. These simulations allow businesses to optimize the use of resources, enhance throughput and production process by modifying configurations in real-time. In addition, Digital Twins of production lines can be used in the food processing industry to optimize its mixing, cooking and packaging processes. As an example, a digital copy of a food producing line can be run through various cooking, ratios of ingredients and level of energy consumption in order to achieve the utmost quality of produced product and minimize energy consumption and waste.

iii. **Real-Time Monitoring**

With the help of Digital Twins, real-time monitoring enables organizations to monitor health and performance of physical systems continuously. This ability becomes priceless in terms of anomaly detection, identification of system states, and proactive reaction to possible problems before they become serious.

How it Works: Digital Twins are linked to the sources of real-time data, IoT devices and sensors, providing an ongoing stream of information on the condition of equipment, systems, or infrastructure. This information is examined to evaluate the performance of the system and raise any anomalies that may be noted.

Example: Transportation systems in smart cities can use Digital Twins to check the traffic flow, the condition of roads, and the work of public transport in them. This information will enable the authorities to optimize traffic lights, re-route buses during peak time, and manage accidents or congestion live, which will facilitate the traffic flow and minimize delays. In addition, in utilities, power grid or water distribution Digital Twins can be utilized to keep track of the health of the system in real-time. Through the performance data of these systems, the operators can identify the inefficiencies or anticipate failure in the infrastructure and repair it in time and thus the risk of disruption of services is minimized [26, 27].

The visual representation of predictive maintenance, process optimization, and real-time monitoring is shown in Fig. 4.2.

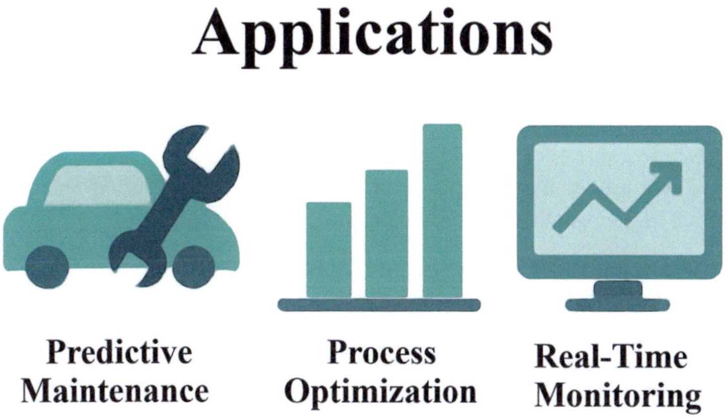

Fig. 4.2 Applications of digital twins

4.5 Advantages of Digital Twins

Adoption of Digital Twins provides diverse benefits in the industries. These advantages have a great influence on the efficiency of operations, quality of the products and customer satisfaction.

i. **Reduced Downtime**

Digital Twins assist in cutting down unexpected downtimes by means of predictive maintenance. Organizations may reduce the effects of system outages by predicting equipment failures and only carrying out maintenance whenever necessary and this will prevent expensive repairs or shutdowns of production.

Example: Digital Twins are applied in the oil and gas sector where downtimes can be very expensive; the pumps, compressors, and other important assets are monitored using Digital Twins. It is possible to predict the failure of these components using predictive analytics that will take the past performance data and sensor values to estimate the likelihood of failure in the future and plan the maintenance and repair during scheduled outages.

Example: In manufacturing, factories can prevent sudden breakdowns caused by wearing, which would paralyze production by monitoring the performance of machines with the help of Digital Twins and identifying the early warning signs. This will make the maintenance to be carried out on time before it causes expensive delays.

ii. **Improved Quality**

Digital Twins also greatly improve quality control since businesses have the opportunity to test the processes and products virtually prior to their actual implementation.

Digital Twins also aid companies in making sure that their products are produced at the highest standards and in case of defects and quality problems, the risk is minimal.

Example: In aerospace, Digital Twins can be applied in the production of aircraft parts as they can test the design of their products and manufacturing process. Simulation of stress, pressure, temperature, etc. allows the companies to detect possible weak points in the components before they are physically manufactured and, therefore, guarantees a higher quality level and safety.

Case study: Digital Twins may be applied in pharmaceutical production as the simulated and optimized production process guarantees the quality of the drugs produced. This involves testing the variables like time of mixing, combinations of ingredients, and settings of the machines to make sure that the end product is tough in terms of regulatory and quality standards.

iii. **Enhanced Customization**

Digital Twins are used to achieve mass customization since it will enable companies to design virtual models of products depending on customer preference or a particular need. The practice will guarantee that companies are able to provide customized services to the clients at an affordable price and with high quality production.

Example: In 3D printing, Digital Twins enable the manufacturers to quickly make changes to the product designs and features according to the feedback or specifications of the customers. Take the case of a customer who desires a custom design on a jewelry, the manufacturer can produce a computer image of the same to capture the exact desires of the customer which they can then proceed with production to produce a near perfect and custom product. In addition, in the fashion industry, Digital Twins can be applied to simulate how the clothes are fitting, what color and what material properties they will have, and designers will be able to customize the clothes to a specific customer even before the actual manufacturing process starts. This leads to less wastage, shortened time-to-market and customer satisfaction [26, 27].

4.6 Problems of Digital Twins Implementation

Though Digital Twins have immense advantages, they are also associated with challenges, which include the quality of the data, compatibility with the current systems, security issues, and implementation cost. The accuracy and reliability of data is also a key to success of Digital Twins and the companies need to invest in high quality infrastructure to process and secure this data. Some of the issues are as follows:

i. **Data Quality and Integration**: It is also important to ensure the data inputted in the Digital Twin is correct and reliable. It is possible to get inaccurate simulations and make informed decisions based on poor-quality data. Also, it might be complicated to combine information on different systems and devices and might need a lot of investment in infrastructure [28–31].

ii. **High Initial Investment**: The cost of Digital Twins implementation, particularly in large operations, entails large upfront expenses that are required to support the needed sensors, IoT systems, and computing resources. The smaller businesses might not be able to afford the required technologies and the competencies.
iii. **Cybersecurity Risks**: Since Digital Twins require real-time data transfer, they may become victims of cyber-attacks. Securing sensitive information and guaranteeing the safety of the connected systems is of great importance in avoiding possible hacking or system crashes.
iv. **Return on Investment (ROI)**: Digital Twins have huge potential in terms of streamlining operations, improving efficiency and decreasing cost. Nonetheless, deployment and scaling of this technology has its challenges especially in the areas of Return on Investment (ROI). Although the value is obvious, the complexities of this challenge should be considered by businesses to the fullest in order to achieve high returns and long-term success.

    ROI Considerations: The ROI of the Digital Twin technology may be tricky at the initial stages since the initial investment is huge. Although the ROI is expected to pay off in the long term, organizations should thoroughly analyze the ROI by considering how the introduction of Digital Twins will lead to the improvement of different operations of their organizations including efficiency of operations, reduction in downtimes, improvement in quality, and overall cost reduction. The companies have to consider the long-term value that Digital Twins will offer to make a clear business case of the investment beyond the immediate cost [32].

    Start-up Costs and Investment: Implementing the Digital Twin technology can be very expensive, especially in sectors which have to deal with large sensor networks, the ability to process a lot of data at a high-performance level as well as complex software integration. To analyze how viable it is to implement Digital Twins by businesses, there is a need to analyze in detail the cost involved in implementing the technology in terms of capital expenditure (CAPEX) and operating expenditure (OPEX).

    Example: In the case of a manufacturing plant, deployment of Digital Twins may involve installing thousands of sensors on equipment, data gathering and analysis software, cloud computing, and professionally trained staff to integrate the system. Such costs may add up to a huge initial outlay before any returns are reaped.

    Measuring ROI: ROI is difficult to measure after the implementation of Digital Twins, though it is necessary to define the success of the technology. ROI can be calculated by dividing the initial investment into the Digital Twin technologies with the savings and performance increase over time.

    Operational Efficiency: Operational efficiency is an area where companies can measure ROI based on how Digital Twins can help to reduce cycle times, more effective resource allocation and accurate demand forecasting etc.

Downtime Reduction: The unplanned downtime can be considerably reduced with the help of predictive maintenance provided by Digital Twins, and the effect of this will be seen in the decrease in operational costs. Businesses can reduce the cost of repairing equipment and minimize revenue loss caused by the breakdown in equipment by preventing equipment failures and optimizing maintenance [33–36].

Quality Improvements: Digital Twins also increase the quality of the products because they constantly monitor the production process and identify its possible deviations in terms of quality. The constant quality helps the businesses minimize wastage, defects and the rework which costs a lot of money.

Example: A power plant that uses Digital Twins to check turbines and other important infrastructure may estimate ROI by dividing the money spent on preventing malfunctions of equipment (e.g., reduced repairs, insurance, and higher reliability) and the original cost of installing the system.

In order to implement Digital Twins successfully, one should pay attention to both ROI and scalability. The cost of the initial investment may be high but the long-term advantages in terms of efficiency, predictive maintenance, and quality improvement can bring a good ROI.

v. **Scalability Challenges**: There are some challenges associated with the expansion of Digital Twin technology in an organization or across multiple facilities, and they are integration with the current systems, data management, and the high cost of expanding the infrastructure.

Compatibility with the current systems: Integration with legacy systems is one of the most important scale-up obstacles of Digital Twins. Most companies have a combination of current digital systems and legacy equipment which might not be readily compatible with the new technologies. To guarantee that the Digital Twin systems could communicate and integrate with existing infrastructure, it is usually necessary to retrofit, upgrade, or replace older systems, which may be time-consuming and expensive.

Example: An old and new production line in a factory: to integrate the Digital Twin of the new production line with the older machines, it might need major changes to the existing machines or even additional sensors to collect the required data. These digital twins might fail to deliver real-time actionable information without an appropriate integration strategy.

Solution: Firms should come up with an integration strategy that does not ignore the firm's existing systems and new technologies. To make the process of integrating legacy systems with modern Digital Twin technology easier, it may be beneficial to invest in middleware or IoT platforms that will allow narrowing the gap between the two systems [37–39].

To address the issue of scalability, the businesses would have to make a transition to integrating the Digital Twin technology into the current systems, dealing with large volumes of data, and possessing the infrastructure to operate globally or in multiple locations.

vi. **Data Management**: With the growing number of assets being digitized and the growing number of sensors, the data generated by the Digital Twins grows exponentially. One of the major challenges in this is how to manage such a huge amount of information, especially in an organization that does not have the infrastructure or expertise to deal with huge data in terms of collection and processing.

    Data Consistency: It is important to know that data is consistent, accurate and reliable. When the data input on a Digital Twin is not accurate or does not match, then the model will give misleading information, and this will result in ineffective decision-making.

    Data Storage and Processing: Companies need to store and process the data, and the amount of data increases, the storage and computing power needed have to be adequate. This, in several occasions, involves investment in higher levels of cloud infrastructure or edge computing to process data near the source and shorten latency.

    Example: In the case of smart cities, when Digital Twins are applied to different elements of the infrastructure, including traffic systems, power grids, and water distribution, the amount of real-time data generated is enormous. Efficient management of this data is only possible with the help of powerful cloud or edge computing infrastructure, and the possibility to process and analyze this data in a short time to get actionable insights [40, 41].

vii. **Scaling Cost**: Although Digital Twins can deliver great value to a single system or asset, it will be expensive to scale the technology to a larger number of facilities or worldwide operations, in terms of infrastructure, cloud computing, and data storage. The expenses involved in increasing the system to support numerous assets, locations, or product lines may be a significant burden to the growth of organizations [42].

    Example: A firm that wishes to expand its Digital Twin usage in one manufacturing facility to an entire supply chain international might have to invest in global cloud systems, high-level data analytics, and the capabilities to support and improve the infrastructure. This must be well planned and invested in ensuring that the technology is scalable.

    Solution: Scalable Digital Twin architecture should be a priority for the companies that want to develop it along with their needs. Taking advantage of cloud-based solutions and modular IoT, it is possible to make the scaling process cheaper and less complex, so that businesses can scale their Digital Twin implementations gradually as they increase in size.

As much as Digital Twins yield instant value in terms of efficiency and minimization of downtimes, it generates long-term value. Continuous improvement can be achieved due to the possibility of constant monitoring, optimization, and adjustment of operations that results in more agile and innovative business processes.

By successfully overcoming these obstacles, businesses will be able to get the maximum out of Digital Twin technology, innovate, make better decisions, and grow sustainably and in the long-term perspective.

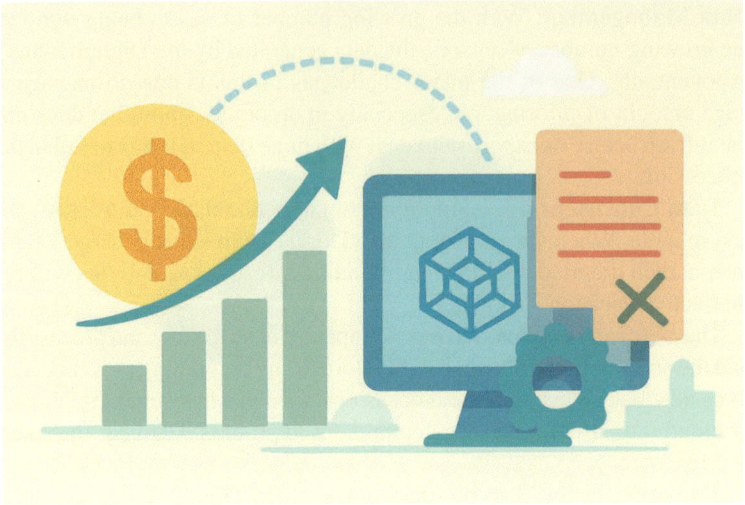

Fig. 4.3 Visual analysis of ROI consideration and scalability challenges

The visual representation of ROI considerations and scalability challenges is shown in Fig. 4.3.

4.7 Learning Outcomes

After going through this chapter, the readers will be in a position to:

- Learn the basic principle of digital twins and architecture, as well as the position of IoT, sensors and real-time analytics.
- Determine the most important areas where digital twins can be used in different industries such as maintenance, optimization of processes, and real-time monitoring.
- Analyze the advantages of digital twins use, which include higher operational efficiency, better quality control, and more options of customization.
- Discuss the issues and factors to consider on implementation and scalability of digital twins especially on cost, integration, and data management.
- Learn the essentiality of digital twins in the development of smart environments that enhance cooperation between human beings and machines and results in increased innovation and sustainability.

4.8 Review Questions

1. Determine the meaning of digital twins and how they work as part of Industry 5.0.
2. (a) How can digital twins help with predictive maintenance in such industries as automotive and aerospace?
   (b) Give examples of how digital twins can minimize the downtimes and maintenance expenses by anticipating the equipment failure before it takes place.
3. (a) Explain how digital twins can be used in the optimization of processes.
   (b) What are the ways through which companies can model, simulate, and optimise their production lines or service processes with digital twins?
4. (a) Describe how digital twins can be used to monitor in real time smart environments, including smart cities, or energy management systems.
   (b) What are the benefits of constant monitoring to improve the efficiency of operations and help in avoiding failure or breakdown of the systems?
5. (a) What are the main advantages of the usage of digital twins in manufacturing sectors?
   (b) Briefly describe how digital twins enhance product quality, product customization and sustainability.
6. (a) Determine the major difficulties that companies can encounter by implementing the concept of digital twins into the current framework.
   (b) What are the scalability problems and legacy system integration and how do they affect the wider use of digital twins in industries?
7. (a) Examine the considerations of the return of investment (ROI) of the implementation of digital twins.
   (b) What are the short term and long-term value propositions companies are likely to receive by investing in digital twin technologies?

Appendix: Solution Manual

1. **Determine the meaning of digital twins and how they work as part of Industry 5.0.**
   Solution: Digital twins are virtual replicas of physical objects or processes, created using real-time data from IoT sensors and devices. These models continuously update and simulate the real-world entity's status. The architecture includes IoT sensors for data collection, cloud or edge computing for data processing, and AI for real-time analytics and decision-making. By maintaining a dynamic feedback loop between the physical and digital counterparts, digital twins enable real-time monitoring and optimization of systems.
2. **(a) How can digital twins help with predictive maintenance in such industries as automotive and aerospace?**

Solution: Digital twins facilitate predictive maintenance by continuously monitoring the health of equipment through sensors. In the automotive industry, a digital twin of a vehicle can track engine performance and detect early signs of wear.

(b) Give examples of how digital twins can minimize the downtimes and maintenance expenses by anticipating the equipment failure before it takes place.

Solution: In aerospace, digital twins of engine components can predict part failures, allowing operators to schedule maintenance before failures disrupt operations, thus reducing downtime and maintenance costs.

3. **(a) Explain how digital twins can be used in the optimization of processes.**

   Solution: Digital twins allow businesses to simulate and optimize various processes in a digital environment before implementation in the real world. For instance, a digital twin of a production line can model different workflows to identify bottlenecks, optimize material flow, and improve resource allocation, ultimately enhancing productivity and reducing waste. In the food processing industry, digital twins help optimize production efficiency by adjusting ingredient mixing and cooking processes in real time.

   (b) What are the ways through which companies can model, simulate, and optimise their production lines or service processes with digital twins?

   Solution: A digital twin of a production line can model different workflows to identify bottlenecks, optimize material flow, and improve resource allocation, ultimately enhancing productivity and reducing waste. In the food processing industry, digital twins help optimize production efficiency by adjusting ingredient mixing and cooking processes in real time.

4. **(a) Describe how digital twins can be used to monitor in real time smart environments, including smart cities, or energy management systems.**

   Solution: Digital twins enable real-time monitoring by continuously tracking the performance of physical systems, such as power grids, traffic management systems, or water distribution networks. For example, in smart cities, digital twins of traffic systems allow for real-time optimization of traffic signals, reducing congestion and improving traffic flow. Similarly, in energy management, digital twins monitor power grid performance, allowing for immediate responses to fluctuations or failures.

   (b) What are the benefits of constant monitoring to improve the efficiency of operations and help in avoiding failure or breakdown of the systems?

   Solution: For example, in smart cities, digital twins of traffic systems allow for real-time optimization of traffic signals, reducing congestion and improving traffic flow. Similarly, in energy management, digital twins monitor power grid performance, allowing for immediate responses to fluctuations or failures.

5. **(a) What are the main advantages of the usage of digital twins in manufacturing sectors?**

Solution: The benefits of digital twins include improved product quality through virtual testing and simulation, enhanced customization by allowing companies to design products based on individual customer preferences, and better sustainability by optimizing production processes and reducing waste.

(b) Briefly describe how digital twins enhance product quality, product customization and sustainability.

Solution: By creating accurate virtual models, digital twins enable companies to anticipate problems and improve product outcomes, all while ensuring cost-efficiency.

6. **(a) Determine the major difficulties that companies can encounter by implementing the concept of digital twins into the current framework.**

   Solution: Integration challenges include the need for extensive infrastructure upgrades to connect digital twins with legacy systems, which may not be designed to handle the real-time data streams from sensors.

   (b) What are the scalability problems and legacy system integration and how do they affect the wider use of digital twins in industries?

   Solution: Scalability is another challenge, as companies may struggle to implement digital twins across multiple locations or production lines without significant investment in IT infrastructure. These issues require careful planning and investment in system integration, cloud computing, and data management.

7. **(a) Examine the considerations of the return of investment (ROI) of the implementation of digital twins.**

   Solution: ROI for digital twins can be evaluated by comparing the initial investment in sensors, cloud infrastructure, and AI analytics with the long-term savings in maintenance costs, improved operational efficiency, and reduced downtime.

   (b) What are the short term and long-term value propositions companies are likely to receive by investing in digital twin technologies?

   Solution: In the short term, companies may see savings through predictive maintenance, while long-term value comes from continuous process improvements, higher product quality, and innovation.

Bibliography

1. Wang Z et al (2022) Mobility digital twin: Concept, architecture, case study, and future challenges. IEEE Internet Things J 9(18):17452–17467
2. Khan WA et al (2025) Industry 5.0. In: Computer aided engineering design and manufacturing: a fourth industrial revolution perspective. Springer, Cham, pp 523–535
3. Khan WA et al (2025) Communication protocols—legacy and Industry 4.0. In: Computer aided engineering design and manufacturing: a fourth industrial revolution perspective. Springer, Cham, pp 485–504
4. Khan WA et al (2025) Nexus of CIM, Industry 4.0, Industry 5.0, and quantum computing. Computer aided engineering design and manufacturing: a fourth industrial revolution perspective. Springer, Cham, pp 551–557

5. Zhong D et al (2023) Overview of predictive maintenance based on digital twin technology. Heliyon 9(4)
6. Johansen SS, Nejad AR (2019) On digital twin condition monitoring approach for drivetrains in marine applications. In: International Conference on offshore mechanics and arctic engineering, vol 58899. American Society of Mechanical Engineers
7. Khan WA et al (2025) Future manufacturing techniques. In: Computer aided engineering design and manufacturing: a fourth industrial revolution perspective. Springer, Cham, pp 459–480
8. Khan WA et al (2025) Standards for CAD/CAE/CAM. In: Computer aided engineering design and manufacturing: a fourth industrial revolution perspective. Springer, Cham, pp 369–395
9. Khan WA et al (2025) Quantum computing. In: Computer aided engineering design and manufacturing: a fourth industrial revolution perspective. Springer, Cham, pp 537–549
10. Akundi A et al (2022) State of Industry 5.0—analysis and identification of current research trends. Appl Syst Innov 5(1):27
11. Rasheed A, San O, Kvamsdal T (2020) Digital twin: values, challenges and enablers from a modeling perspective. IEEE Access 8:21980–22012
12. Tao F et al (2018) Digital twin in industry: state-of-the-art. IEEE Trans Industr Inf 15(4):2405–2415
13. Hammad M, Islam MS, Salam MA, Jafry AT, Ali I, Khan WA (2023) Framework for the implementation of smart manufacturing systems: a case in point. Processes 11(5):1436
14. Yu W et al (2017) A survey on the edge computing for the Internet of Things. IEEE Access 6:6900–6919
15. Yousefpour A et al (2019) All one needs to know about fog computing and related edge computing paradigms: a complete survey. J Syst Architect 98:289–330
16. Khan WA et al (2025) Computer-aided numerical analysis techniques. In: Computer aided engineering design and manufacturing: a fourth industrial revolution perspective. Springer, Cham, pp 305–355
17. Khan WA et al (2025) Engineering design. In: Computer aided engineering design and manufacturing: a fourth industrial revolution perspective. Springer, Cham, pp 11–63
18. Hammad M et al (2025) Lightweight and robust key agreement for securing IIoT-driven flexible manufacturing systems. IEEE Internet Things J
19. Khan WA et al (2025) Use of databases for CAD/CAE/CAM/CIM. In: Computer aided engineering design and manufacturing: a fourth industrial revolution perspective. Springer, Cham, pp 397–407
20. Xu X et al (2021) Industry 4.0 and Industry 5.0—inception, conception and perception. J Manuf Syst 61:530–535
21. Boschert S, Rosen R (2016) Digital twinning simulation aspect. In: Mechatronic futures: challenges and solutions for mechatronic systems and their designers. Springer, Cham, pp 59–74
22. Hammad M et al (2023) A provable secure and efficient authentication framework for smart manufacturing industry. IEEE Access 11:67626–67639
23. Khan WA et al (2025) Manufacturing and production. In: Computer aided engineering design and manufacturing: a fourth industrial revolution perspective. Springer, Cham, pp 65–185
24. Ghobakhloo M et al (2023) Behind the definition of Industry 5.0: a systematic review of technologies, principles, components, and values. J Ind Prod Eng 40(6):432–447
25. Maddikunta PKR et al (2022) Industry 5.0: a survey on enabling technologies and potential applications. J Indus Inform Integr 26:100257
26. Grieves M, Vickers J (2016) Digital twin: Mitigating unpredictable, undesirable emergent behavior in complex systems. In: Transdisciplinary perspectives on complex systems: New findings and approaches. Springer International Publishing, Cham, pp 85–113
27. He C et al (2019) Time-frequency domain turbo equalization for single-carrier underwater acoustic communications. IEEE Access 7:73324–73335
28. Glaessgen E, Stargel D (2012) The digital twin paradigm for future NASA and US Air Force vehicles. In: 53rd AIAA/ASME/ASCE/AHS/ASC structures, structural dynamics and materials conference 20th AIAA/ASME/AHS adaptive structures conference 14th AIAA

29. Khan WA et al (2025) Production planning and control. In: Computer aided engineering design and manufacturing: a fourth industrial revolution perspective. Springer, Cham, pp 425–453
30. Khan WA et al (2025) Remanufacturing. In: Computer aided engineering design and manufacturing: a fourth industrial revolution perspective. Springer, Cham, pp 187–211
31. Khan WA et al (2025) Mathematics for CAD/CAE/CAM. In: Computer aided engineering design and manufacturing: a fourth industrial revolution perspective. Springer, Cham, pp 357–368
32. Negri E, Fumagalli L, Macchi M (2017) A review of the roles of digital twin in CPS-based production systems. Procedia Manuf 11:939–948
33. Hammad M et al (2025) Revolutionizing aerospace manufacturing with Industry 4.0. In: Smart manufacturing blueprint: navigating Industry 4.0 across diverse sectors. Springer, Cham, pp 87–113
34. Khan WA et al (2025) Pillars of Industry 4.0 and CAD/CAM. In: computer aided engineering design and manufacturing: a fourth industrial revolution perspective. Springer, Cham, pp 509–522
35. Botín-Sanabria DM et al (2022) Digital twin technology challenges and applications: a comprehensive review. Remote Sens 14(6):1335
36. Karunaratne T et al (2025) Evaluating the economic impact of digital twinning in the AEC industry: a systematic review. Buildings 15(14):2583
37. Fuller A et al (2020) Digital twin: enabling technologies, challenges and open research. IEEE Access 8:108952–108971
38. Tao F et al (2018) Digital twin-driven product design, manufacturing and service with big data. Int J Adv Manuf Technol 94(9):3563–3576
39. Hammad M et al (2025) Framework for implementing smart manufacturing. In: Smart manufacturing blueprint: navigating Industry 4.0 across diverse sectors. Springer, Cham, pp 19–44
40. Jones D et al (2020) Characterising the digital twin: a systematic literature review. CIRP J Manuf Sci Technol 29:36–52
41. Lu Y et al (2020) Digital Twin-driven smart manufacturing: connotation, reference model, applications and research issues. Robot Comput Integr Manuf 61:101837
42. Duan H et al (2025) The development of a digital twin concept system. Digital Twin 2530298

Part III
Human-Centricity and Workforce Transformation

Chapter 5
Human-Centric Manufacturing and Workforce Transformation

5.1 Objectives

- To discuss the evolution of workforce functions, which are nowadays based on manual work to management and innovative problem-solving in the framework of Industry 5.0.
- To realize how the combination of collaborative robots, AI, and IoT is transforming job roles and improving the cooperation between people and machines.
- To examine the significance of ergonomic design and safety in the design of human-friendly workspaces.
- To debate upskilling, lifelong learning and vocational training as the essential elements in equipping the workforce to the challenges of Industry 5.0.
- To analyze the ethical concerns related to AI such as bias, privacy and inclusivity and how these concerns can be resolved in Industry 5.0.
- To evaluate workforce transition models and their effects on employment patterns, employment generation and labor market agility against technological dislocation.

5.2 Scope

With the advent of Industry 5.0, the approach to the manufacturing process is radically transformed because the workforce becomes the focus of technological changes. The role of collaborative robots (cobots), artificial intelligence (AI) and the Internet of Things (IoT) in the manufacturing process introduces a radical change in how workspaces, job positions and employee well-being are treated. Unlike in the past industrial revolutions which were labor-intensive and involved mechanization, Industry 5.0 is all about human interaction with machines which makes the workers central to innovation and manufacturing. This part discusses how the workforce is changing, and how the changes are reshaping the nature of work, including the shift

M. Hammad et al., *Industry 5.0: The Human-Centric Future of Industry*, Springer Series in Advanced Manufacturing, https://doi.org/10.1007/978-3-032-12622-1_5

in the workforce, where it is no longer based on manual labour, but the creation of oversight, creative problem-solving, and strategic functions.

5.3 Shifting Positions of the Workforce: Manual Work to Control and Innovation

Industry 5.0 is transforming the way employees relate with the manufacturing processes. Whereas the previous industrial revolutions, including Industry 4.0, were focused on efficiency improvement due to automation and mechanization, Industry 5.0 is based on a more human-friendly approach, in which the role of workers is not to be substituted by machines but to cooperate with them. The work is no longer manual and repetitive, and now it needs cognitive skills, creativity and strategic thinking.

5.3.1 Change in the Role of the Workforce

- **Manual Labor**: Manual labor was the pillar of production in the early years of industrialization. The employees had to work on the assembly lines carrying out repetitive, physical duties. These were mostly low-skill, labor-intensive and usually did not need too much mental work. With Industrial Revolution, most of the laborers performed manual jobs to encourage mass production and satisfy increasing demand.

  Examples: In the traditional manufacturing settings, the assembly line workers would be engaged in routine activities like fixing the parts or checking the parts with little scope of decision-making or creativity. These were physically tiring and mostly repetitive jobs that did not provide employees with much scope of expanding or participation.
- **Automation and Transition to Oversight**: As Industry 4.0 emerged, much of these routine duties started being automated and replaced by robots, which lowered the number of jobs that required human labour. This tendency does not stop in Industry 5.0, and instead of displacing employees, the accent is made on supplementing human functions. Automated systems are now being monitored and managed by workers, and it comes with a new set of skills such as problem-solving skills, system monitoring skills, and trouble shooting skills when problems occur.

  Example: Robots in an Industry 5.0 manufacturing facility can be used to complete the monotonous work of assembling, packaging or inspecting quality. Nonetheless, robotic systems are now under the supervision of human workers who have to maintain efficient functioning of the systems, real-time adjustment of the processes, and solving mechanical and operational problems. The worker has changed his role in carrying out the tasks by handling and optimizing automated systems.

AI-Driven Monitoring: The workers can also monitor the AI-powered systems that forecast and avoid production delays, keep track of the equipment performance, and suggest changes to the process through the analysis of real-time data. This change demands that the workforce be more technically competent and data-insight-reading so that they can maintain the smooth functioning of the processes and be able to solve the problems in a short period of time.

- **Creative and Strategic Roles**: Industry 5.0 will emphasize a lot on human creativity and strategic thinking which machines are effective in terms of efficiency, but they cannot match human skills in terms of ingenuity and emotional intelligence. With robots and AI replacing the more menial and routine jobs, human employees are being asked to pay more attention to problem-solving, design thinking, innovation and value-added activities, which demand a creative approach.

  Example: In the motor industry, robots can be able to assemble the parts of a car, but human beings are needed in product designing and customization. Employees can be required to create new features, come up with a new version of the product depending on the customer demands, or enhance the appearance of the product. Machines are able to do highly structured tasks as compared to human beings, but they are unable to do tasks such as creativity and the capability to solve multifaceted problems.

  Customization and Personalization: As the concept of mass customization has emerged in manufacturing, human workers are required to customize products to suit individual specifications of customers. They might leave the production to AI and cobots, but the creative input will have to be done by humans as they will determine whether the products will satisfy the specifications regarding personalized design [1].

  Example: Customization in high end consumer electronics, based on color, functionality and unique features may need human control and feedback in design stage. Although machines do most of the work, human workers play an important role in making sure that the end product satisfies the customer in terms of needs and expectations.

This Fig. 5.1 categorizes roles in the modern workforce into three distinct areas: manual labor, represented by a person handling boxes; automation, shown with a worker and a robotic arm; and creative and strategic roles, represented by a person using a laptop and thinking. It highlights the evolving nature of work and the shift towards a balance of physical and intellectual tasks.

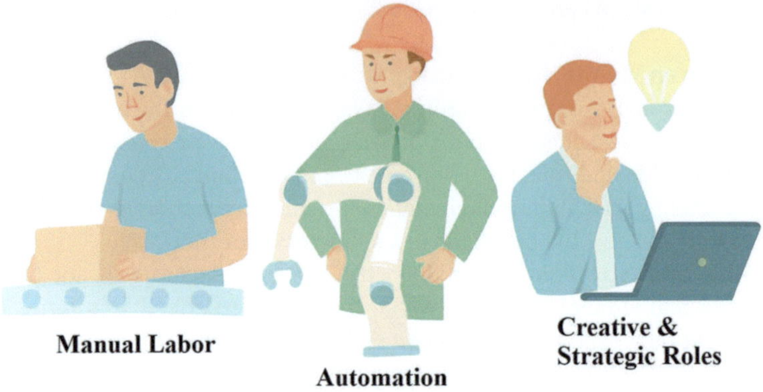

Fig. 5.1 The role of manual labor, automation, and creative roles in workforces

5.3.2 Key Benefits

The reshaping of the job positions in Industry 5.0 has a number of great benefits to the employees and the companies. The transformation of manual work into supervision, innovations, and strategic planning is not only good for employees in respect of participation and working satisfaction, but also leads to improved production, innovation, and business [2].

- **Increased Participation**: With the change in job description whereby the demanding manual work is replaced with the more dynamic supervisory and innovative work, employees take a more active part in the process of production. The capacity to add meaningful input, involvement in the decision-making process and the determination of the production outcome makes one satisfied and motivated with the job. An active workforce will also tend to invest in their work, and this will result in increased output and job satisfaction.

  Example: An employee who monitors a collaborative robot (cobot) in a smart factory will have a higher probability of having a sense of ownership and responsibility for the successful work of the production line. The worker can have a say in the efficiency of the system, and this makes him more involved in the process.
- **Job Satisfaction**: When workers get engaged in meaningful activities that depend on their skills and talents, problem solving, and creativity, they have more chances to be satisfied with their jobs. With the focus on human-related positions, the industry 5.0 allows the employees to concentrate on the spheres where they can really excel and enjoy the work more and feel more satisfied with the results.

  Example: Staff members assigned to the product development teams, where they are able to brainstorm, experiment with the designs, and add their creativity to the process of product development, will be more satisfied than doing the routine assembly line tasks. The change of roles to more strategic makes the workers feel that they have a say in the development of the final product.

- **Improved Productivity**: Automation will help ensure that a balance between efficiency and innovation is met by the business through the combination of automation and human control and creativity. Automation will help simplify the routine processes, whereas people will dedicate their attention to solving complex tasks and innovation. Such an interplay of human creativity and machine efficiency results in increased productivity, better innovation and development of superior products.

  An example is a factory with cobots taking care of the repetitive tasks of assembly, allowing the human workers to work on the production method, propose new features on the design, or address the bottlenecks in the manufacturing process. Consequently, efficiency is gained, and the production process is streamlined and efficient [3].
- **Innovation**: As human beings take charge of strategic decisions and operations control, the businesses are more likely to be innovative. Employees are able to use creativity in order to enhance the process, come up with new product lines as well as come up with innovative solutions to problems and this increases the competitive edge of the company in the market.

5.4 Ergonomics and Safety: Creating the Workplace that is Human Friendly and Machine Friendly

Industry 5.0 is not only about implementing such advanced technologies as AI, robotization, and automation in the manufacturing industry but also about making sure that the human workforce is the focus of the innovations. The most important part of this man-oriented strategy is the safety and welfare of the employees in the process of cooperating with machines. Workspaces have to be designed in a way that focuses on ergonomics, safety, and health to minimize injuries in a workplace, increase productivity, and create general satisfaction among workers. With man and the machines interacting more and more, it is necessary now more than ever that a considerate design is used that would assist in physical and mental wellbeing [4].

5.4.1 Design of Human Machine Workspaces

- **Ergonomic Principles**: In Industry 5.0, cobots (collaborative robots) and automated systems become more and more integrated into the production environment, and the workstations have to be designed in a way that would suit both human workers and machines. The ergonomic principles guarantee that such workspaces are made more comfortable for the workers to minimize the level of strain and chances of injury.

  Heights adjustable and Motion friendly Layouts: The workstations should be adjustable to different body types and work. An example is the assembly lines where the desks or workstations can be adjustable in height to fit the person in the line so that the work is done at an optimum ergonomic angle thus minimizing the physical stress on the back, neck and arms.

Non-Repetitive: In traditional manufacturing, employees used to engage in repetitive movements that may cause musculoskeletal disorders (MSDs). The design of industry 5.0 should be in such a way that it limits repetitive actions and promotes diverse movements to avoid physical distress. Workstations might be made to enable the workers to switch between standing and sitting so that they move dynamically and feel less fatigued.

Example: In automated factory, an assembly line might also include ergonomic workstations that change depending on the requirements of the worker. Tables that could be adjusted, padded floors, and automated tools that fit human ergonomics could assist workers in avoiding physical stress at the end of the long shift.

- **Safe Human–Robot Interaction**: As Industry 5.0 focuses on the collaboration of human and robots, it is crucial to maintain safe interactions between a human worker and a machine. Cobots are developed to collaborate with humans and do not substitute them, and their safe introduction into the working area is a very important issue.

  Force Sensors and Collision Detection: Cobots can also be fitted with sensors that show whether a human being is near them. These sensors enable the robots to halt or reduce their speed instantly in the case they collide with a worker and this eliminates accidents. Detection of collision and force-limiting characteristics make sure that robots do not hurt people when they are around them.

  Safety Monitoring and System Updates: The monitoring of safety should be done regularly such that the cobots remain within the safe limits. The workplaces in Industry 5.0 have to be fitted with real-time safety monitoring systems that monitor the movement of robots, proximity of the workers, and working conditions. It is also important that the software of the cobots should be updated regularly in order to make them safer and more functional.

  Example: In the case of warehouse, cobots may be deployed to move the goods with human workers. The robots can automatically stop or change the path in case a worker steps in its way with advanced collision detection, which avoids any physical injury.
- **Health monitoring with wearable Technology**: Wearable technology is crucial in enhancing physical health and safety of the workers in Industry 5.0. Such gadgets track the health of the worker in real-time and give an instant response to the worker and the management.

  Exoskeletons: Exoskeletons may assist the workers with extra strength during heavy lifting and lower the possibility of musculoskeletal injuries. Such wearable technologies assist the joints and muscles of the body and make workers lift heavy objects without overstraining their body.

  Smart Glasses: Smart glasses are examples of wearables that can be used to give augmented reality (AR) that will show the real time status of the machine, safety measures or assemblage instructions which will reduce the cognitive load

Fig. 5.2 Workplace ergonomics and automation integration

and enhance accuracy. Also, smart glasses are able to warn the workers about the safety hazards in real-time.

- **Health Monitoring Wearable Technology**: Smartwatches or fitness trackers that are installed in the workplace can check the heart rate and fatigue of workers as well as their general health condition. Such wearables remind employees when they should take a break or change the position to prevent fatigue to decrease the probability of injury and increase their output [5].

  Example: An employee wearing a smart exoskeleton during an assembly line operation can get real-time information on his or her posture and muscle load that will enable him or her to adjust to minimize fatigue and injury.

Figure 5.2 shows a combination of human-centered ergonomics and automation at the workplace. It features a robotic arm positioned above a desk while a worker sits at the desk. The worker's posture is emphasized, highlighting the ergonomic setup. The image also includes icons related to safety and work guidelines.

5.4.2 Benefits

- **Less Workplace Injury**: Ergonomic work design with a strong focus on the interaction between people and machines can be effective in minimizing the chances of getting work-related injuries, especially musculoskeletal disorders (MSDs) and injuries caused by repetitive work.
- **Better Comfort of the Worker**: Designing workstations ergonomically will increase the comfort level that will result in better job satisfaction and reduced

absenteeism. The workers will tend to remain involved in their work and will contribute well to the work culture when they are comfortable and safe in their surroundings [6].

- **Improved Productivity**: A healthy and comfortable working environment limits the distractions and physical discomfort of the workers which makes them concentrate better on the job. Reducing the chances of injury and fatigue will ensure that the companies can have a higher level of productivity and operational efficiency [7].

5.5 Careers and Competencies: Upskilling, Lifelong Learning and Vocational Training

With Industry 5.0, the workspace and manufacturing space are changing more than ever before, and this means that new skills are more needed than ever. The upskill of the workforce needs to be approached holistically because the shift to more cognitive, creative and strategic jobs is a trend. Vocational training and lifelong learning are also essential to make sure that the workers would be able to fulfill the changing requirements of the industries and industries of the future and would not be replaced by automated robots.

5.5.1 Industry 5.0 Upskilling

- **Training of the Current Employees**: The possibility of providing the existing workforce with the skills needed to cooperate with new technologies such as AI, robotics, and IoT is one of the primary challenges of Industry 5.0. Companies should engage in upskilling programs to make their workers prepared for the digital transformation.

  Example: A manufacturing company may introduce training facilities to educate the workers on the techniques of monitoring and controlling cobots, managing production lines powered by AI, and solving technical problems that occur with automation systems. These programs are tailorable to enable the employees to acquire technical skills and be transferred to more involved positions.
- **Lifelong Learning**: Due to the rapidly increasing rate of technological changes, employees have to be ready to engage in life-long learning in order to stay competitive. Industry 5.0 promotes lifelong learning initiatives where the employees should be able to update their knowledge and skills to meet the new tools, processes and trends in the industry [8].

  Example: Online courses or in-house training might provide the latest in the field of robotics programming, integration of AI, and data analysis so that the employees could keep up with the time and advance their careers without having to switch employers.

Fig. 5.3 Professional growth, learning, and automation in the workplace

- **Vocational Training**: Vocational training is important in equipping the workers to perform specialized jobs in the industry 5.0. As we continue to embrace the use of more complicated systems such as robotics, artificial intelligence, and cyber-physical systems, vocational training will ensure that the workers acquire the technical knowledge needed in the new jobs [9].

  Example: Vocational education in the field of robotics programming, AI system management, and machine learning can teach the current generation of employees to work in a profession that is only a few years old.

Figure 5.3 emphasizes professional development in the context of automation and technology. It features a person working at a desk with a laptop, surrounded by symbols representing growth, learning, and automation, including a robotic arm and a staircase symbolizing career advancement.

5.5.2 Benefits

- **Enhanced Employee Flexibility**: Employees that are subjected to constant training and upskilling become more flexible to deal with disruptions that can be caused by technology and changes in industry needs. Such flexibility will make sure that employees are up to date and can be of use in making Industry 5.0 successful.
- **Higher Job Security**: When employees are equipped with current skills, it is less likely that they will be replaced by automation. With retraining and the development of new skills, employees will be able to shift to higher levels of work, which

will go hand in hand with automation, guaranteeing employment in the long-term perspective.

- **Accelerated Technology Adoption**: Incorporation of new technologies in daily activities can be done within a short period by a well-trained workforce, which enables firms to realize quick adoption of Industry 5.0 technologies. This increases productivity, better innovation, and the digital transformation process becomes easy [10].

5.6 Ethical Concerns: AI Bias, Privacy, and Inclusivity

The incorporation of AI, IoT, and other smart systems in the labor force is changing the modes of operation of businesses in Industry 5.0. But along with such technological development, there are also a series of ethical issues that need to be considered properly. As the AI is becoming an increasingly large part of the decision-making process and the amount of personal data being gathered is also increasing, the ethical issues of AI algorithm bias, data privacy, and inclusiveness in the workplace become central. The above is some of the challenges that industry 5.0 will have to deal with, so that technological development benefits all the stakeholders through fairness, security, and equality.

5.6.1 Ethical Issues of Industry 5.0

- **Discrimination in AI**: The AI systems are meant to go through large data and learn on it and thus they are the best tools that can improve efficiency and accuracy as well as decision making. Nevertheless, if the dataset that AI algorithms are trained on is biased, it may cause unintended discriminatory effects and reinforce the current biases in society. This is especially alarming in such fields as hiring, promotion, law enforcement, and healthcare where such decisions by AI can increase disparities.

  Example: AI algorithms are frequently applied in the process of selecting job candidates. In case the training data is biased in the past, i.e., it supports the candidates of a particular demographic group over others, then the algorithm will discriminate against applicants on the basis of race, gender, or socioeconomic background. This may result in discriminatory employment policies that breach diversity and equality of opportunity in the labor market.

  Reducing Bias: This is one of the primary measures to reduce bias in AI to make sure that the data that is utilized to train such systems is varied, representative and not affected by the historical bias. Corporations need to focus on the development of datasets that will represent a wide range of demographic groups so that the AI systems can treat everyone equally.

Example: In the medical sector, discriminatory AI might result in unfair recommendations of treatment towards various races or socioeconomic groups. Auditing and correcting biased data will help healthcare providers to make equity-based decisions using AI, lowering the disparity of care.

- **Privacy Concerns**: Privacy is a major concern in the 5.0 industry as more and more IoT devices, wearable technologies, and big data are used. Technology gathers a tremendous amount of personal data from workers, consumers and activities. The possibility of data leaking, spying, and unwarranted access to confidential information becomes a big problem. Businesses should introduce powerful data safeguards to protect personal data and honor the right to privacy of individuals.

  Example: Smartwatches or exoskeletons, which are worn in the manufacturing setting to track health and performance of workers can gather sensitive information concerning the physical condition of an employee, their movements, or even their emotions. In case this data is improperly processed, employees will suspect that their privacy is being infringed which results in the issues of surveillance and loss of autonomy.

  Data Security: To deal with the issue of privacy, organizations are required to have strong cybersecurity systems to help them safeguard personal information against theft or misuse. This is through encrypting sensitive information, anonymizing personal information where feasible and ensuring that the data is only accessible by authorized personnel only.

  Example: In a smart factory, the information collected by IoT sensors that monitor the work of machines and employees should be secure so that personal data of employees were not disclosed. Also, the companies are advised to practice transparency by telling the employees and the consumers about the nature of the data collected, its usage and the people who will access it.

- **Inclusiveness**: Since the workforce is being transformed with technological changes, it is important that the changes are made to be inclusive and not to deepen the existing inequalities. Industry 5.0 should also be able to guarantee the inclusion of marginalized groups (women, racial minorities, and people with disabilities) in the gains of the technological advancement. To have a just and fair workplace, it is important to develop inclusive practices that will encourage diversity and maintain equal opportunities for everyone.

  Example: The problem of lack of women and people of color in technical jobs is persistent in the tech business. Depending on the approach to the workforce development and diversity, Industry 5.0 can either contribute to the continuation of these disparities or become the force that introduces more inclusive practices.

  Enhancing inclusivity: Companies can enhance inclusivity through the development of accessible workplaces, equal opportunity of training and career development and active promotion of diverse work teams. Also, organizations can adopt policies that would make emerging technology available to every employee including the ones having physical disability by making the workplace systems to include assistive technologies and universal design features [11].

  Example: At an AI development firm, one should strive to make sure that the AI systems that are created are not only created to represent the values of a

Fig. 5.4 Diversity, ethics, and digital privacy in the workplace

homogeneous group of engineers but are also open to various cultural, social, and ethical views. It is possible to do this by engaging various teams in the design and testing of AI algorithms.

Figure 5.4 represents the intersection of workplace ethics, privacy, and diversity. It includes various icons related to security and ethics, alongside images of diverse individuals, emphasizing the importance of inclusivity and ethical considerations in the digital age.

5.6.2 Solutions

The solution to the ethical issues of Industry 5.0 is of great value to the organizations and society in general. With the focus on justice, confidentiality, and inclusion, companies will be able to establish a more just and reliable atmosphere, and this will eventually result in the long-term success of the business.

- **More Equitable Working Conditions**: Industry 5.0 can make workplaces fairer by focusing on such aspects as bias in AI and inclusivity. A just working place implies that every employee irrespective of his or her background has equal chances of succeeding and prospering. The result is better morale, increased employee participation and a more diversified and innovative working population.

Example: An organization which removes the bias in the AI-based hiring system will make sure that the candidates are recruited depending on their experience and potential, but not on their demographic features. This leads to a more diverse and skilled workforce.

- **Enhanced Trust**: By using ethical AI, keeping privacy and inclusivity in mind, companies establish good relations with their employees, consumers, and investors. Trust is crucial to long-term relations and brand loyalty. Consumers and employees would be willing to buy the products of organizations that pay much attention to ethical issues and prove their interest in fairness and privacy.

  Example: A company which is open about its data collection policy, making sure that AI algorithms are not biased in any way, and making an effort to spread diversity and inclusion will earn the trust of its customers. Such trust may result in more customer loyalty, good word-of-mouth and better brand reputation.
- **Social Responsibility**: The ethics of Industry 5.0 helps in fostering social responsibility as companies should engage in practices that are good for business as well as society. By tackling the issues of bias, privacy and inclusivity, businesses are able to play a role towards wider social aims of equality, fairness and responsible technology usage. These activities improve the image of a company and make it fit into the social values, which is getting more and more important to both consumers and investors [12].

  Example: An inclusive AI-building company, privacy-sensitive to its users, and a company that values diversity among its employees is contributing to the well-being of society by alleviating the systemic inequalities and encouraging the responsible technological development.

5.7 Transition Models of the Workforce and the Effects on the Employment Trends

With more and more industries turning to automation, artificial intelligence (AI) and collaborative robotics, the transformation of the workforce is an absolute necessity to allow seamless transition and reduce the adverse effect on the employment pattern. The new technologies of industry 5.0 are in essence transforming the labor market, integrating the capabilities of humans and the advanced technologies, which poses some challenges and opportunities to the workers. To make the most out of automation and reduce the job displacement as well as encouraging long-term economic development, models of effective workforce transition are required.

5.7.1 Transition Models of Workforce

- **Retraining and redeployment**: Retraining the workers to give them skills that they need to perform in new job positions is one of the most important elements of a successful workforce transition. With automation, AI and robotics replacing manual repetitive jobs, a large number of employees will have to work in other fields where human capabilities, including creativity, problem-solving, and decision-making, are still required.

Oversight and Technical Training: The employees who were engaged in manual labor in the past might require some training to monitor the automated system, control the AI processes, or program and maintain robotics. As an example, the workers may switch the assembly line work to the functions in AI management, robotics programming, or system monitoring. The jobs also need an in-depth knowledge of technology, how systems work and how problems can be found and resolved.

Continuous Learning Programs: Investment in continuous learning programs is needed by the companies to make sure that their employees are flexible enough to change with the emerging technologies. Such programs may contain workshops, certifications, and online courses on such skills as data analysis, machine learning, and software development.

Example: A manufacturing firm which has brought robotic automation in its production line can offer training to its employees on how to maintain, program and supervise the robots so that the workers are able to handle and operate the robots effectively and solve any technical problems that might occur.

- **Collaborative Workflows**: The anthropocentric approach of industry 5.0 promotes the development of working processes where humans and AI, robotics, among other technologies collaborate to enhance their abilities. As opposed to the traditional models whereby automation exposes humans to redundancy, Industry 5.0 creates an atmosphere in which technology and human intelligence collaborate in the process of maximizing productivity.

  Dynamic Roles: The roles of employees will also become dynamic and multi-faceted as the employees work with AI and robotics. They will have to be technically skilled as well as innovative in thinking to improve the functioning of AI systems and robotics, solve problems, and make strategic choices.

  Example: A high-tech manufacturing plant can have humans collaborating with collaborative robots (cobots) to manage production, ensure quality and personalize the products. As robots work on repetitive tasks, the employees add innovative thinking to design thinking, decision-making and process enhancement.

  Flexible and Agile workforce: Collaborative workflows also allow more flexible and agile workforce structure where employees can be flexible to different roles and activities according to the project requirements. This will encourage cross-functional collaboration as employees will cooperate in the process of problem solving and innovation with AI and automation systems.
- **Creation of New Jobs**: Although the automation and AI will likely eliminate certain jobs, Industry 5.0 will generate new ones, particularly in fields where there is a need to monitor, develop creative solutions, and handle technical matters. These emerging occupations will play a fundamental role in future of work.

New Job categories: The job categories that will be created are in the field of AI management, cybersecurity, data analysis, and high-tech manufacturing. The current trends of more businesses becoming dependent on AI and robotics means that there will be an increase in the demand of professionals who are able to handle such technologies. The employees will be expected to learn to deal with complicated systems, meet the challenge of cybersecurity, and make sure that robots and AI work effectively and ethically.

Green Industries: Green economy is another industry that is set to develop in Industry 5.0 since it is concerned with eco-friendly activities and sustainability. With sustainability emerging as a more significant concern, there will be more jobs in renewable energy, waste disposal, and sustainable product development [13].

Example: As AI and automation in the logistics industry continues to expand, new positions will be created, including the position of AI systems coordinator or a robotics supervisor, who will be in charge of the process of integrating autonomous fleets and warehouses. In the same way, there will be new employment opportunities in the area of sustainable technology development and environmental policy as the companies move to greener practices.

Figure 5.5 depicts an educational setting where industrial professionals are being taught about AI and automation. One worker is reading a manual, while a lecturer points to an image of a robotic arm. The setting suggests a learning environment focused on the integration of new technologies in the workforce.

Fig. 5.5 AI and automation in industrial education and training

5.7.2 Effect on Employment Patterns

- **Workplace Destruction and Technical Creation**: Although automation is poised to eliminate certain manual jobs, it is expected to generate even more new and more valuable jobs whose tasks and responsibilities are centered on creativity, monitoring, and teamwork. Policymakers need to be conversant with these changes and put in place measures to curb job loss and enhance generation of new jobs.

  Job Displacement: Routine jobs that are easily mechanized especially the manual ones, will most probably be taken over by AI and robots. These involve positions in production, storage and simple customer service. Nevertheless, it is possible that a great number of these jobs will be transformed into more technical or supervising, instead of being cut off completely.

  New categories of jobs: Automation, on the other hand, will introduce new job categories, especially those in the fields of AI management, data analysis, cybersecurity, and the green industries. There will be more demand for jobs that involve creativity, emotional intelligence and problem solving as human beings supplement the machines.

  Example: Although some jobs on the assembly line can be replaced with automation, new jobs in the field of machine maintenance, work on robotics, and optimization of systems will emerge, and this way, the workforce will still be occupied with valuable and productive work.
- **Remote and Flexible working**: The digital technologies will provide the possibility to speed up the changes in remote and flexible working, which will be particularly encouraged by Industry 5.0. With an increased level of digitization and automation of industries, employees will be able to work more freely and flexibly.

  More Remote Work: Employees in industries like software development, customer services, and project management will spend more time working remotely and will not require physical offices, and they will also save time by not commuting to work as a result of the emergence of digital tools.

  Enhanced Work-Life Balance: Teleworking may provide employees with a healthier work-life balance since they are free to create their own work hours and work at places that are convenient for them. This flexibility is especially attractive to workers that might have care giving responsibilities or any other personal commitments.

  Example: A group of data analysts assigned to a global project can be separated by several time zones and locations, and the implementation of collaboration tools in the cloud and AI-based data analysis will enable them to work in any place, making the work more comfortable and efficient.
- **Greater Diversity in the workforce**: Industry 5.0 is also likely to make workforce more diverse, especially since AI and automation will eliminate physical barriers to work. Increased women and disabled people as well as other marginalized groups will have a chance to enter the workforce and make the job market more inclusive and equitable.

Diversified Work Practices: AI and automation can be used to make workplaces more inclusive by offering assistive technologies, i.e., voice-controlled systems, ergonomic tools, and AI-based adaptive devices. Such innovations will ease the ability of people with disabilities to play roles that were previously inaccessible [14].

Example: In a warehouse, heavy lifting can be done with automated systems and thus employees with physical disability can do jobs that they could not do before. Moreover, the option of flexibility in the work environment allows people who might have been discriminated against in the conventional workplaces to excel in their new positions.

5.7.3 Benefits

- **Less loss of jobs**: Industry 5.0 can minimize the adverse effects of automation on work by paying attention to retraining, upskilling, and redeployment. Instead of losing their jobs, workers can shift to other new jobs that supplement the technological changes.
- **Better Labor Markets**: Labor markets will be strengthened through the establishment of new categories of the job and the possibility to develop skills. Workers will become more flexible and able to withstand changes that technology has brought about as they are provided with the skills to be used in the emerging sectors.
- **Innovation, Productivity and Economic Growth**: An innovative, productive, and economically developed country will be based on a well-developed workforce that is capable of changing job positions. With the new industries and new technologies employed by businesses, the economy as a whole will gain the benefits of the additional value provided by adaptable and skilled workforce [15].

5.8 Learning Outcomes

After the completion of this chapter, the readers will be able to:

- Understand how the role of workforce has changed in terms of manual labor to more creative and strategic workforce and the role of human in Industry 5.0.
- Assess the use of more modern technologies such as cobots, artificial intelligence, and internet of things, and how they can be used to make the manufacturing process more human decision-making and creative.
- Examine the impact of ergonomics and safety in the preservation of the health of workers in human–machine collaborative settings.
- Explain the role of upskilling, lifelong learning and vocational training in equipping the workforce with emerging jobs in Industry 5.0.
- List the ethical issues of AI linked to the algorithmic bias and data privacy and suggest ways to achieve workplace fairness and inclusiveness.

- Understand the workforce transition models and the implications for job creation, displacement, and the future of labor markets in the context of Industry 5.0.

5.9 Review Questions

1. (a) What are the impacts of industry 5.0 that change the job of the human worker and replace it with more strategic and creative duties?
   (b) Briefly describe the changes in the roles of the workforce since previous industrial revolutions and the new demands of workers in Industry 5.0.
2. (a) What are the main advantages and difficulties of collaborative robots (cobots) implementation into the manufacturing environment?
   (b) Examine how cobots can be used to improve productivity, safety, and flexibility and how to deal with the possible issue of losing jobs.
3. (a) How does Industry 5.0 and AI and machine learning enhance human decision-making?
   (b) Give concrete examples of how AI can support predictive analytics and real-time decision-making and make sure that human creativity and judgment are focused.
4. What are the ways of making workspaces safe to the workers especially in a place where man and machines are in close contact?
5. (a) How does Industry 5.0 meet the requirement of upskilling the workforce?
   (b) Explain the role of lifelong learning, vocational training, and constant education programs in equipping the worker to take up jobs in the field of human–machine collaboration and the new industries.
6. (a) Which are the ethical issues of AI and data privacy in Industry 5.0?
   (b) What can be done to reduce these problems so that AI systems are fair, inclusive and transparent?
   (c) Give examples of avoiding AI bias and how companies can preserve the privacy of the workers.
7. (a) What impact will the trend of automation and AI in the manufacturing industry have on job displacement and what are the new jobs likely to appear because of it?
   (b) Examine how reskilling and retraining will help to make a smooth transition in the workforce.
8. In what ways are workforce transition models in Industry 5.0 different to those in other industrial revolutions?
9. How might Industry 5.0 foster inclusivity in the workplace in terms of staff diversity?
10. (a) What is the part of green industries and sustainability in the workforce transformation of Industry 5.0?

(b) What are some of the ways in which the move towards becoming more eco-friendly opens up new employment opportunities and what are some of the skills required of the employees in these new fields?

Appendix: Solution Manual

1. **(a) What are the impacts of industry 5.0 that change the job of the human worker and replace it with more strategic and creative duties?**

   Solution: Industry 5.0 shifts the role of workers from manual, repetitive tasks to positions requiring creative thinking and decision-making. While previous industrial revolutions emphasized manual labor, the human-centric focus of Industry 5.0 encourages workers to engage in more strategic, oversight, and innovative roles.

   (b) Briefly describe the changes in the roles of the workforce since previous industrial revolutions and the new demands of workers in Industry 5.0.

   Solution: Workers now contribute to areas such as product development, problem-solving, and design thinking, which require human creativity and critical judgment.
2. **(a) What are the main advantages and difficulties of collaborative robots (cobots) implementation into the manufacturing environment?**

   Solution: Collaborative robots (cobots) offer several benefits, including increased productivity through task assistance, enhanced safety by preventing worker injuries, and flexibility in reprogramming to perform different tasks.

   (b) Examine how cobots can be used to improve productivity, safety, and flexibility and how to deal with the possible issue of losing jobs.

   Solution: However, challenges include the potential for job displacement, especially in roles that were once physically demanding but now require oversight. Cobots can help mitigate this by enhancing the worker's role rather than replacing it entirely.
3. **(a) How does Industry 5.0 and AI and machine learning enhance human decision-making?**

   Solution: AI and machine learning augment human decision-making by providing predictive analytics and data-driven insights that help inform business decisions.

   (b) Give concrete examples of how AI can support predictive analytics and real-time decision-making and make sure that human creativity and judgment are focused.

   Solution: For example, AI can predict maintenance needs and suggest improvements in operational efficiency. The human worker remains responsible for final decisions, ensuring that creativity, ethical considerations, and emotional intelligence remain integral to the decision-making process.

4. **What are the ways of making workspaces safe to the workers especially in a place where man and machines are in close contact?**

   Solution: Designing workspaces for human–machine collaboration requires ergonomic principles such as adjustable workstations and motion-friendly layouts. Wearable technologies like exoskeletons and smart glasses help monitor workers' physical conditions, alerting them to risks such as fatigue or improper posture. These designs reduce the risk of musculoskeletal injuries and ensure that both humans and machines can work in harmony, promoting safety and comfort.

5. **(a) How does Industry 5.0 meet the requirement of upskilling the workforce?**

   Solution: Upskilling is essential in Industry 5.0, as workers must be trained to manage and collaborate with emerging technologies.

   (b) Explain the role of lifelong learning, vocational training, and constant education programs in equipping the worker to take up jobs in the field of human–machine collaboration and the new industries.

   Solution: Lifelong learning programs that focus on robotics, AI interaction, and data analytics ensure that workers remain adaptable. Vocational training, particularly in specialized fields such as AI management and cyber-physical systems, prepares workers for the evolving technological landscape.

6. **(a) Which are the ethical issues of AI and data privacy in Industry 5.0?**

   Solution: Ethical concerns include the risk of AI algorithms reinforcing societal biases and issues related to data privacy.

   (b) What can be done to reduce these problems so that AI systems are fair, inclusive and transparent?

   Solution: Mitigating these issues involves using diverse datasets for AI training to minimize bias and implementing strict data protection policies.

   (c) Give examples of avoiding AI bias and how companies can preserve the privacy of the workers.

   Solution: Ensuring transparency in AI decision-making processes and protecting personal information will build trust and create a more equitable and secure workplace.

7. **(a) What impact will the trend of automation and AI in the manufacturing industry have on job displacement and what are the new jobs likely to appear because of it?**

   Solution: Industry 5.0 will lead to both job displacement and creation. While automation may eliminate some manual roles, new jobs in AI management, robotics programming, and creative problem-solving will emerge.

   (b) Examine how reskilling and retraining will help to make a smooth transition in the workforce.

   Solution: The need for reskilling and retraining programs is critical to ensuring that displaced workers can transition into these new roles and contribute to the innovative processes of Industry 5.0.

8. **In what ways are workforce transition models in Industry 5.0 different to those in other industrial revolutions?**

Solution: Workforce transition models in Industry 5.0 focus on retraining and redeployment to ensure that workers can adapt to technological changes. Collaborative workflows, where human creativity complements machine efficiency, are central to this model. This transformation ensures that workers are involved in strategic and creative tasks, fostering a more dynamic and productive work environment.

9. **How might Industry 5.0 foster inclusivity in the workplace in terms of staff diversity?**

   Solution: Industry 5.0 promotes inclusivity by removing physical barriers to work, allowing marginalized groups to participate more fully in the workforce. AI and automation can enable remote work, flexible schedules, and assistive technologies that support diverse abilities, helping to create a more inclusive and equitable workplace.

10. (a) What is the part of green industries **and sustainability in the workforce transformation of Industry 5.0?**

    Solution: Green industries and sustainability play a significant role in creating new job opportunities within Industry 5.0.

    (b) What are some of the ways in which the move towards becoming more eco-friendly opens up new employment opportunities and what are some of the skills required of the employees in these new fields?

    Solution: Roles related to eco-friendly practices, such as sustainability consultants and green technology specialists, are emerging. Workers in these sectors need skills in environmental management, renewable energy systems, and sustainable manufacturing practices to contribute to a more sustainable future.

Bibliography

1. Khan WA et al (2025) Computer aided engineering design and manufacturing: a fourth industrial revolution perspective. Springer
2. Khan WA et al (2025) Industry 5.0. In: Computer aided engineering design and manufacturing: a fourth industrial revolution perspective. Springer, Cham, pp 523–535
3. Xu X et al (2021) Industry 4.0 and Industry 5.0—inception, conception and perception. J Manuf Syst 61:530–535
4. Trstenjak M et al (2025) Human factors and ergonomics in industry 5.0—a systematic literature review. Appl Sci 15(4):2123
5. Hammad M et al (2023) Framework for the implementation of smart manufacturing systems: a case in point. Processes 11(5):1436
6. Tóth A et al (102260) The human-centric Industry 5.0 collaboration architecture. Methods X(11), 102260
7. Trstenjak M et al (2023) Cognitive ergonomics in industry 5.0. In: International conference on flexible automation and intelligent manufacturing. Springer, Cham
8. Katuk N (2024) The future of human-computer integration
9. Olcott E (2024) Robot revolution' forces Chinas human workforce to adapt. Financial Times, 14 Nov 2024.

10. Naeini S, Hassan GMC, Mosaddad SH (2022) Industrial design evolution in the context of ergonomics and industry 5.0. J Design Thinking 3(2), 165–172
11. Khan WA et al (2025) Nexus of CIM, Industry 4.0, Industry 5.0, and quantum computing. Computer aided engineering design and manufacturing: a fourth industrial revolution perspective. Springer, Cham, pp 551–557
12. Liang C-J et al (2024) Ethics of artificial intelligence and robotics in the architecture, engineering, and construction industry. Autom Construct 162:105369
13. Khan WA et al (2025) Communication protocols—legacy and Industry 4.0. In: Computer aided engineering design and manufacturing: a fourth industrial revolution perspective. Springer, Cham, pp 485–504
14. Jha S, Bag D (2019) The service sector: migration, technology and productivity. Int J Sociol Soc Policy 39(1/2):2–21
15. Toth I (2021) Thriving in modern knowledge work: personal resources and challenging job demands as drivers for engagement at work

Part IV
Sustainability and Circular Economy

Chapter 6
Sustainability and Circular Economy in Industry 5.0

6.1 Objectives

- To discuss the ways of incorporating sustainability into the very fiber of Industry 5.0 such as energy efficiency, waste reduction, and emissions control.
- To discuss how the circular economy can transform the process of manufacturing, concentrating on the issues of reuse, remanufacturing, and recycling.
- To examine the cost-saving, regulatory compliance, and brand reputation as the benefits of sustainability in Industry 5.0.
- To determine how the industry 5.0 technologies assist the companies in achieving the more and more strict environmental requirements and international regulations.
- To learn why performance measures (KPIs) and data-driven decision-making in sustainability is important and can enhance environmental performance.
- In order to determine whether Life Cycle Assessment (LCA) and Cost–Benefit Analysis (CBA) is useful in making sound judgments concerning sustainable practices.

6.2 Scope

With the world struggling to deal with the looming environmental crises, the principles of sustainability and circular economy are being centralized in the industrial practices. Industry 5.0 acknowledges that it is high time to balance the development of industries and the environment, and it is necessary to make sure that new technologies do not only boost the development of the economy but also make the planet and its people healthier. This part will discuss the implementation of sustainability into Industry 5.0 through energy efficiency, minimization of waste, and control

M. Hammad et al., *Industry 5.0: The Human-Centric Future of Industry*, Springer Series in Advanced Manufacturing, https://doi.org/10.1007/978-3-032-12622-1_6

of emissions, and the discussion of circular economy in determining the future of manufacturing.

6.3 Sustainability as a Core Driver

Sustainability in Industry 5.0 is not a mere add-on or an appendix to the business but a primary force of innovation, operation planning and sustainability. With businesses trying to reduce their environmental impact and at the same time satisfy the demands of consumers and make more profits, their interest is in increasing energy efficiency, waste reduction, and emission control in their production processes. Companies incorporating these pillars of sustainability into their business are contributing to the development of a more responsible industrial system that would be favorable both to business and the environment [1, 2].

6.3.1 Energy Efficiency

Sustainable manufacturing entails energy efficiency since industries consume a huge percentage of energy in the world. Industry 5.0 strives to implement smart energy management systems that are based on the IoT devices, AI, and machine learning to maximize energy consumption in real-time.

Smart Energy Management Systems: With the help of advanced IoT sensors and AI-based solutions, Industry 5.0 provides an opportunity to monitor the use of energy in manufacturing facilities. These systems gather real-time information about the use of energy, the functioning of the machines, and environmental factors to help businesses make the most of the energy they use and prevent unnecessary losses of energy. As an example, AI algorithms can automatically switch the lighting, heating, and cooling systems depending on occupancy and weather conditions, thus not using the energy when it is not needed.

Predictive Energy Management: AI-based systems have the ability to forecast energy demand trends on the basis of past experiences, seasonality and production schedules. This would allow the manufacturers to regulate operations in advance and minimize energy use during peak demand times or when the energy prices are high. Reduced energy wastage means that companies can drastically cut their running expenses and contribute less to environmental pollution.

Renewable Energy: Industry 5.0 promotes the use of renewable energy sources in industries including solar, wind and hydropower. Using solar panels, wind turbines, and other renewable energy options in manufacturing plants will help industries to decrease their dependence on fossil fuel and minimize the emission of carbon. A good number of firms are already using solar powered systems in powering production lines, warehouses and office buildings and this has greatly improved the carbon intensity of their operations.

Energy Storage Solutions: Energy storage solutions can also assist businesses in storing any renewable energy produced during peak energy production hours to use when there is low energy supply. This does not only make energy more resilient but also makes it less dependent on outside power sources, which makes the energy infrastructure more sustainable and independent [3].

6.3.2 *Waste Reduction*

The reduction of waste is another major principle of Industry 5.0, and the final aim is to reach zero-waste production. Industry 5.0 is all about reducing waste throughout the entire chain of the production process, including the process of obtaining raw materials and delivery of the final product.

Zero-Waste Manufacturing: Zero-waste manufacturing implies the design of production that will not produce waste at all. This may involve redesigning the raw materials utilization, streamlining production processes and recycling the by-products into the production cycle. Industry 5.0 will make use of intelligent systems that evaluate the efficiency of production, use of materials, and minimize wastage of resources.

Closed-Loop Manufacturing Systems: Closed-loop system is developed so that materials and by-products of the production process are used continuously. Businesses recycle and reuse materials instead of sending them to the landfills, and thus they make sure that the materials are put back into the manufacturing process. As an illustration, in the electronic sector, the manufacturers can reuse defective or idle parts to make new products and minimize new raw materials.

AI and Machine Learning of Prevention of Waste: AI and machine learning algorithms are being more and more used to improve the production line and minimize waste. The given technologies will have the capabilities to analyze the data in real-time within production systems and identify the inefficiencies, overproduction, faulty products, or the overuse of materials. Such AI systems can forecast the potential problems in advance thereby eliminating wastage even before they occur.

Supply Chain Optimization: This is another Industry 5.0 concern, which involves reduction of the waste in the entire supply chain. Introduced as a result of the use of digital twins, AI-powered logistics, and real-time tracking, the companies are able to control inventory more effectively, optimize the transportation process, and reduce the volume of packaging waste. The streamlined supply chain will ensure that the products are shipped in an efficient and least disruptive manner [4].

6.3.3 Emissions Control

This is because the manufacturing industry is also among the major sources of carbon emission and other pollutants, hence reducing emissions is an important element of the sustainability approach of Industry 5.0. This can be achieved through the integration of clean technologies, emission control and energy saving machines.

Carbon Capture and Storage (CCS): Carbon capture and storage (CCS) is among the most likely technologies of the emissions control. Carbon dioxide (CO_2) released into the atmosphere by industrial practices is not allowed to penetrate into the atmosphere but is trapped in the CCS systems. These systems are implementable in the manufacturing facility, power stations and other heavy industries to reduce their carbon footprint by a large margin.

Real-Time Emissions Monitoring: Industry 5.0 will use AI and IoT to check the emission in real-time that provides businesses with real-time data on their performance regarding the environment. With the continuous monitoring the companies will be in a position to notice the sources of over emissions and make corrections on the production processes. Complex algorithms can also help manufacturers to regulate their energy use and produce the minimal amount of emissions changing the temperature, pressure, and speed of the machinery on the fly.

Electric Vehicles and Machinery: Industry 5.0 also promotes the use of electric machines and vehicles (EVs) in the manufacturing plant, other than reducing the number of emissions produced in the process of manufacturing. Firms can minimise the amount of greenhouse gases emitted through the replacement of fossil-fuel powered vehicles and equipments with electric powered vehicles and equipments. This is more so when it comes to automotive and logistics sectors where their conventional vehicles and machinery are primary contributors to emissions [5].

Figure 6.1 is the sustainability in Industry 5.0, which deals with energy efficiency, reduction of waste, and control of the emissions. It brings to the fore use of new technologies to optimize energy use and incorporation of renewable energy sources such as solar and wind power, and the trend toward zero-waste production. The clean technologies applied in the emission control include carbon capture and production technologies using electric vehicles. These are in a bid to minimize impact of industrial activities on the environment.

6.3.4 Benefits

There are numerous advantages that come with the incorporation of sustainability in Industry 5.0, not only in the environment, but also to businesses, consumers and to society in general.

- **Cost Savings**: Improving energy efficiency, reducing waste, and incorporating renewable sources of energy would make businesses cut down on the cost of

Fig. 6.1 Industry 5.0 with innovations in energy efficiency, waste reduction, and emissions control

operations. Sustainable manufacturing practices also save the use of raw materials, reduce disposal costs and result in saving energy and other resources in the long run.

- **Regulatory Compliance**: Implementation of sustainability initiatives assists companies to remain in line with the strict environmental policies. New laws are being enforced by the governments of different countries to reduce emissions, to minimize wastage as well as to preserve the environment. The firms that will be proactive in adopting sustainability will be well placed to comply with these requirements as well as escape penalties.
- **Reputation**: More consumers are relying on sustainability as a source of differentiation. Business organizations which place priority on environmental responsibility and practice sustainability improve their image and create customer loyalty. The businesses that adhere to their values have higher chances of gaining the support of the consumers, and as a result, their market share and profitability will grow [6].

6.4 Industrial Processes with Embedded Circular Economy Principles

Circular economy is one of the essential elements of Industry 5.0, which questions the relevance of the linear cycle of production and consumption, which is often referred to as the take-make-dispose one. It encourages a regenerative model in which resources are used until they are no longer of use, in which the greatest value can be obtained out of them, and waste is reduced. Implementing ideas of reuse, remanufacturing, and recycling in industrial activities, Industry 5.0 will allow creating an industrial ecosystem that will not lead to environmental degradation and will instead become more sustainable and resilient. This part discusses the incorporation of such principles of the circular economy in Industry 5.0, paying attention to the most important processes of reuse, remanufacturing, and recycling, and pointing out the environmental and economic advantages that they are followed by.

6.4.1 Reuse

Reuse is one of the fundamental concepts of the circular economy, and this concept is dedicated to the necessity to prolong the life of products and components instead of their destruction. Industry 5.0 focuses on designing and operating the process that makes reuse and reconditioning of products easier and does not require new resources.

Design, Modularity and Repairability: Industry 5.0 promotes the products to be designed in a modular and repairable fashion. They can easily replace, upgrade or reuse the components of modular designs; this increases the life span of the whole product. This design philosophy is in contrast with the traditional products that are single-use and where it is necessary to discard the whole product when just one component breaks.

Examples: In consumer electronics, firms are able to make smartphones whose batteries and other parts are easily replaceable. Rather than having to get rid of an entire phone once a single part breaks, now the end user can just get the broken part replaced, leading to less e-waste and the extension of product lifecycles.

In the automotive sector, cars can be made in such a way that one can easily change the major parts of the vehicle including the engine, transmission, and tires without having to change the whole car. This will not only make repairs to be more affordable but also it will mean that less resources will be required to replace products that have not run out of usable parts.

Product-as-a-service (PaaS) Models: Product-as-a-Service is getting momentum especially in segments like electronics, appliances, and transportation. As opposed to selling the products, companies provide them on subscription or leasing basis. This makes sure that the products are returned to the manufacturer when they are at the end of their lifecycle so that they can be refurbished, repaired or sold. The model focuses on reuse of product by maintaining it in the market longer thus requiring less production of new products [7].

Example: In the office equipment, there are such businesses as Xerox that provide printers and copiers for leasing. When the lease expires, the equipment is returned, repaired and resold, so that it keeps on being used and the waste is reduced as well.

Industry 5.0 adopts these reuse principles and in the process, it minimizes the need to consume raw materials, decreases the pressure on landfills, and establishes a more effective and sustainable method of consuming goods.

6.4.2 Remanufacturing

Remanufacturing is the process of renewing the used products to a like-new condition or even replacing worn parts with new parts. This cycle has the advantage of prolonging the life of products, less demand for raw materials and less environmental degradation than that of making new products. Industry 5.0 focuses on the use of remanufacturing technology to reduce wastage and resources utilization [7].

Repairing of Product Value: Remanufacturing is an environmentally friendly option for new manufacturing, which dwells upon reusing the existing products and returning them to the state which is either as good as the first or even better. The remanufacturing operations tend to be more energy efficient than the production of new parts and less raw materials are needed.

Examples: In automotive industry parts such as engines, transmissions and chassis may be remanufactured to the original factory specifications. Taking them apart, cleaning, and replacing the worn-out components with new ones, companies can bring the product to the state of the new one, saving the raw materials and the energy needed to recreate the same components.

Remanufacturing is also adopted in the electronics industry whereby the older products, including smart phones, computers, and printers are repaired and upgraded to be sold or redistributed. It does not only minimize the amount of e-waste but also lowers the environmental costs of producing completely new devices.

6.4.2.1 Advantages of Remanufacturing

There are many benefits of remanufacturing in terms of economy and environment:

- **Less Resource Depletion**: The need for virgin raw materials is also saved through reuse of components and this saves the natural resources.
- **Energy Savings**: Energy consumption is usually lower in the process of remanufacturing than production of new products hence leading to reduction of carbon emission and operational costs.
- **Economic Opportunity**: The Remanufacturing industry provides employment opportunities through repair, refurbishing and quality control and thus develops the local economy [8].

6.4.3 Recycling

The other important value of the circular economy is recycling, whereby the useful materials in the products that have been discarded are reused and are re-introduced into the production chains. Industry 5.0 introduces new technologies for recycling materials to enhance material recovery, waste reduction and the use of virgin materials.

High-technology Recycling: Industry 5.0 uses technologies like AI, robotics, and machine learning in automation and optimization of the process of sorting, recycling, and recovery of materials. The technologies help to enhance the efficiency of the recycling process and thus materials like metals, plastics, and textiles are re-used in the manufacturing industry, and not disposed of landfills.

AI-Powered Sorting systems: AI-powered sorting systems are used in the accurate sorting of recyclable materials e.g. metals, plastics, and glass, to be separated within the waste streams. These systems enhance efficiency of recycling plants, contamination and the quality of the recycled materials by automating the sorting process.

Example: In the automotive sector, the old cars (end-of-life vehicles (ELVs) may be recycled through the application of novel shredding and sorting technologies. The valuable components, namely, aluminum and steel, can be separated with the help of AI and robotics and recycled, thus being reused in the production of new cars along with plastic and rubber parts.

Closed-Loop Recycling: Closed-loop recycling means that the recycling of material is done into the same type of product and thus the use of virgin material will be reduced. This is particularly necessary in industries such as electronics whereby the demand of the rare earth metals can be mitigated through recycling of the available materials [9].

Examples: In the electronics sector, closed-loop recycling allows the manufacturers to recover the precious metals like gold, silver and copper used in the old products and recycle them to make new products. This saves a lot on the aspect of mining which is a resource intensive and environmentally destructive process.

In the textile sector, closed-loop recycling allows recycling fibers of old clothing into new clothes and decreases the waste problem of textiles on the environment.

Figure 6.2 visualizes three key components of a circular economy in Industry 5.0: reuse, remanufacturing, and recycling. It highlights how products, such as electronics and motors, can be reused or remanufactured rather than discarded. The recycling symbols surrounding these items represent the closed-loop system encouraged in Industry 5.0, where materials are continuously recycled and reprocessed to reduce environmental impact and resource consumption.

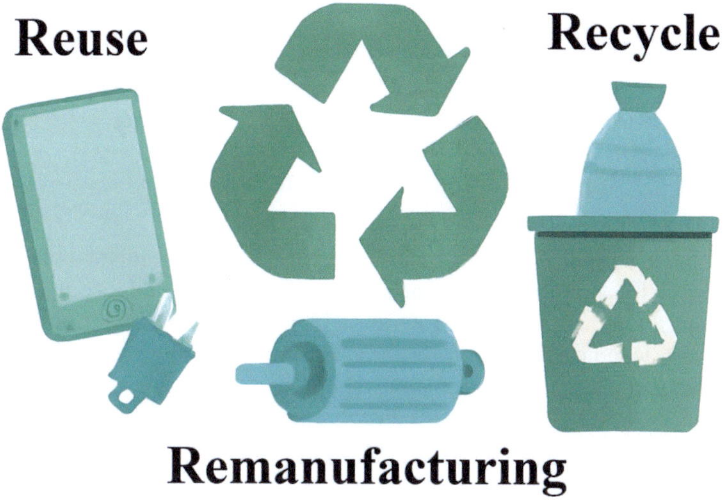

Fig. 6.2 Industry 5.0 emphasizing the power of recycling, reuse, and remanufacturing

6.4.4 Benefits

The inclusion of the circular economic principles in Industry 5.0 is associated with numerous advantages, which go beyond environmental sustainability. Companies are able to use reuse, remanufacturing and recycling in industrial practices and thus achieve a lot of resource savings, cost savings and development of more sustainable business models.

- **Less consumption of resources**: Industry 5.0 contributes to the reduction of the pressure on extracting and processing raw materials by focusing on reusing, remanufacturing, and recycling materials, most of which are resource-hungry and harmful to the environment. This change in the conventional take, make, dispose model is beneficial in the conservation of natural resources including metals, plastics, and water so that they can be used in future.
- **Sustainable product Design**: The principles of circular design guarantee the design of products that can be made to last long, repaired and recycled. This does not only improve sustainability of products in their life cycle but also minimizes the necessity to produce new products all the time. Through sustainable product design, the businesses will be able to design products that can be maintained easily, repaired and recycled, which will help in reducing the amount of waste.
- **Reduced Environment Footprint**: Businesses can also reduce the impact on the environment significantly by incorporating the principles of circular economy in the manufacturing processes. Reuse of materials, remanufacturing of goods and the best use of recycling cut pollution, save energy and cut down on the environmental impact of industrial activities. Such initiatives allow reducing the negative

impact of conventional manufacturing processes and build a more sustainable future [10].

6.5 Regulatory Landscape: Environmental Standard and Global Agreements Compliance

Industry 5.0 is a highly dynamic environment in which the regulations in the world are trying to focus more on sustainability and environmental accountability. The governments, international organizations and the consumers are putting pressure on the industries to reduce their effects on the environment, employ cleaner technologies, and comply with the rigid environmental regulations. As the demand grows on industries to make sure that their operations do not contradict the environmental regulations and international obligations, Industry 5.0 can provide the technology solutions that would support the process. This section serves to investigate the regulatory context comprising of national environmental standards, international agreements such as the Paris Agreement, and how Industry 5.0 can help business to comply with the regulations through better technologies.

6.5.1 *Environmental Standards*

The nations in the globe are developing more rigid environmental laws that will limit pollution, minimize wastes and conserve natural resources. These legislations address the most important environmental issues such as carbon emission, water consumption, waste disposal and chemicals utilized in production. Industry 5.0 will help businesses to monitor, control, and report on their environmental performance and this is in compliance with those standards with the assistance of AI, IoT, and automation as the advanced technologies.

Carbon Emission and Climate Change laws: Minimization of carbon emission is among the topmost environmental concerns in the industries of today. The governments are implementing stricter policies in regard to carbon emission of the industrial activities in a bid to curb climate change. It is now mandatory that companies take up technologies which assist them to measure, reduce and mitigate their emissions.

Examples: The Emissions Trading System (ETS) of the European Union and other carbon trading programs in other parts of the world compel the companies to monitor and report their carbon emissions. Such Industry 5.0 technologies like IoT-based monitoring systems and AI-based analytics can help the manufacturers to track emissions in real-time, detect inefficiency in the manufacturing processes and take corrective actions to minimize their carbon footprint.

AI-based emissions monitoring systems would be able to examine its energy consumption, processes of production, and equipment performance to propose operational adjustments that would save energy and eventually reduce carbon emissions.

Waste management and Water Usage: The industries that use quite a lot of water resources, i.e., textiles, food processing, and manufacturing industries are now regulated and can use only a certain amount of water and release certain amounts of waste. The industries must also treat and handle the wastewater in an effective manner to avoid water body pollution.

Examples: In textile industry, processes that consume a lot of water such as dyeing and washing have been subjected to review because of water scarcity issues. Industry 5.0 proposes the use of IoT sensors and automated water management systems that can track the usage of water in real-time and offer the companies data that can reduce the wastage of water, recycle water, and comply with the regulations.

AI algorithms can be applied to optimize the waste management process in the manufacturing industry, automatically sorting the waste materials into the recycling process and to make sure the waste disposal process is in line with the local and international regulations.

Hazardous materials and Chemical Use: The manufacturing industry is not an exception to regulations on the use of dangerous chemicals. As an example, international agreements limit the application of toxic substances (including lead and mercury and some solvents). Industry 5.0 provides remedies to enable the companies to lessen the use of hazardous chemicals through incorporating green chemistry, cleaner production and safer substitutes in their manufacturing procedures [11].

Example: During the manufacture of semiconductors, hazardous chemicals are used but their use is controlled because of their possible effect on the environment. The development of Industry 5.0 technologies will allow automating the process of handling, storing, and disposing of chemicals, thus making sure that all the operations are conducted safely and in line with the environmental laws.

6.5.2 International Agreements and Observance

International treaties like the Paris agreement have established high ambitions that the countries should work towards reducing the emission of greenhouse gases and curbing global warming. Such agreements not only force governments to act but also businesses because they are forced to become more sustainable. Industry 5.0 is a key factor in ensuring that companies are able to meet international agreements since it supplies the required technology to track and minimize emission, energy use, and waste.

Paris Agreement: The Paris Agreement, in which almost 200 countries participate, is to ensure that the rise in world temperature is well below 2 °C above the pre-industrial levels, and an aspiration is that the rise should be less than 1.5 °C. To attain this objective, countries are making efforts to mitigate carbon emission, switch to renewable energy as well as adopt more sustainable industrial activities. Although the Paris Agreement is mainly aimed at governments, there is a lot of responsibility laid on industries to ensure that they minimize their carbon footprints.

Paris Agreement Industry 5.0 Solutions: To achieve the Paris agreement, companies have to track and cut down their emissions. Industry 5.0 makes this possible by offering solutions like tracking emissions in real-time, smart energy management, and combining renewable sources of energy into the manufacturing operations. Through the implementation of these technologies, businesses will have a chance to reduce national and global carbon emission goals and streamline the use of energy in their firms.

Example: A manufacturing company can rely on renewable sources of energy, such as solar or wind energy, with the aid of AI systems, to maximize energy consumption and minimize dependence on fossil fuels, which will contribute to achieving the emission reduction targets provided in the Paris Agreement.

SDGs Sustainable Development Goals (SDGs): Sustainable Development Goals (SDGs) offered by the United Nations offer a universal framework of reaching environmental sustainability, eradicating poverty, and social equality. Some of the SDGs, including Goal 12 (Responsible Consumption and Production) and Goal 13 (Climate Action), have a direct connection to the aims of Industry 5.0, which makes it less challenging to help these goals to be achieved by the businesses through their sustainable practices.

Industry 5.0 and SDGs: The emphasis on the principle of circular economy in Industry 5.0, which includes reuse, remanufacturing, and recycling, assists companies in the achievement of SDG 12 that aims to minimize waste, conserve resources, and establish sustainable production patterns. Also, the use of renewable energy sources and emission control technology enhances goal 13, which assists the industries in achieving the climate action goals [12].

Figure 6.3 depicts a visual representation of Industry 5.0's role in ensuring compliance with both national and international environmental standards. It features a checklist representing environmental regulations and the Paris Agreement logo, indicating global efforts to reduce carbon emissions and promote sustainable practices.

6.5.3 Advantages of Adherence to the Environment Standards and Global Agreements

Complying with the environmental rules and conventions has many benefits to businesses such as the evasion of fines, global competitiveness, and reputation. Such

Fig. 6.3 Compliance with environmental standards in the age of Industry 5.0

advantages allow businesses to be sustainable in the long term, as well as support the sustainability trend in the world.

- **Avoidance of Penalties**: Failure to comply with the environmental standards may lead to hefty fines, prosecution, or sanction that will make the business difficult to run. These penalties can be avoided by ensuring that the companies are in compliance with the national and international regulations to reduce chances of legal and financial penalties.

  The fines imposed by the European Union on the companies that exceed the limits of carbon emissions as established by the Emissions Trading System (ETS) provide an example. With the help of the modern monitoring systems offered by the 5.0 technologies, the companies will be able to remain within the lines of the existing regulations and prevent such costly fines.
- **Global Competitiveness**: Businesses that meet international environmental standards enjoy competitive advantage as sustainability becomes a priority of consumers, investors and stakeholders. Firms which are complying with or surpassing the environmental requirements are more appealing to the consumers and investors who are focused on sustainability, they are more likely to enjoy a good relationship with the government and regulatory authorities.

  Example: A business which follows renewable energy, minimizes its carbon footprint, and applies the principles of circular economy will be able to stand out in the market as an environmentally friendly enterprise, winning the affection of socially responsible consumers and eco-investors.
- **Branding and reputation**: Environmental regulation and agreements with the world have a great contribution in making a company have a good reputation.

There is a growing pressure on businesses to be responsible towards the environment as demanded by the consumers. The image of a company can be enhanced by its adherence to fewer emissions, less waste and saving resources which will bring loyal customers and enhance the market position of the company [13].

Example: Tesla and Patagonia companies have managed to create their brands on the basis of being sustainable. They have developed a good reputation by observing the environmental standards and consequently this has led to their success in the market.

6.6 Measurement and Reporting: KPIs of Sustainability Performance

In order to spur up the sustainability agenda, companies should not only embrace sustainable business, but they should also be in a position to quantify their environmental performance effectively. This will need a strong framework of monitoring sustainability objectives, making changes where necessary and ensuring that there is an improvement. Industry 5.0 focuses on the significance of Key Performance Indicators (KPIs) and data-based insights, which can be crucial in measuring sustainability initiatives of different industries. With the help of the most cutting-edge technologies including IoT, AI and automation, companies can incorporate the use of real-time data collection and analysis into the sustainability measurement activities, which then may result in more effective decisions and increased environmental performance.

6.6.1 Sustainability KPIs

KPIs play a crucial role in monitoring the sustainability initiatives of any business, and they can assist the company by providing a clear direction, monitoring the progress, and measuring the overall effect. Industry 5.0 promotes the embracement of superior technologies in taking key sustainability measures that can enable companies to take data-driven decisions that can support environmental objectives.

- **Energy Efficiency**: Sustainable manufacturing and operating are characteristics of energy efficiency. Energy efficiency KPIs assist one to track energy usage in an organization and areas that need improvement, which leads to cost reduction and environmental friendliness.

  Energy Consumption per Unit of Production: This KPI monitors the quantity of the energy consumed per unit of production which allows determining the level of energy efficiency of the manufacturing processes. Through this data, companies have an opportunity to detect the inefficiencies and streamline production to cut down energy waste.

Energy Costs Reduction: The KPI is a percentage change in energy costs due to the increase in energy efficiency. Monitoring this measure assists companies to determine the success of energy conservation efforts and technologies.

Renewable Energy Utilization: Most of the companies are striving to shift to the use of renewable energy to use it to run their operations. This KPI monitors the amount of energy that is generated using renewable sources (solar, wind, or hydroelectric) against the conventional fossil fuels. The higher the percentages, the more commitment to the reduction of carbon footprints and sustainability.

Example: A manufacturing plant can measure the amount of energy that is used per unit of production so as to know the amount of energy needed to produce a unit of product. The plant can maximize energy consumption and minimize expenses and achieve sustainability goals by incorporating AI-based energy management systems.

- **Waste Reduction**: Another important KPI that should be considered by the businesses that want to embrace the principles of the circular economy is the reduction of waste. Businesses are now considering ways of reducing waste production, recycling materials and reusing components to cut their burden on the environment.

  Waste Recycled: This KPI monitors the volume of the waste that is diverted to landfills and is recycled. It assists companies to monitor their progress in attaining zero-waste targets and sees to it that more of their waste is being re-utilized effectively.

  Percentage of Materials Reused: The KPI approximates the ratio of the raw materials, components or packaging used during the production process that are reused. The reduction in the reliance of the business on virgin materials can be achieved by increasing the number of the materials used that are reused hence reducing the effects on the environment.

  Waste Reduction Via Automation and Process Optimization: This is a KPI that quantifies the amounts of waste that are reduced through automation and other process optimization programs. Through the assistance of AI and IoT-based optimization tools, the businesses will be able to minimize the amount of waste generated in the production process and optimize the resources overall.

  Example: A food processing company can track the reused or recycled packaging materials within its production chain. To reduce the waste of the environment, the company can enhance the packaging designs and adopt sequencing systems using AI to reduce the effects.

- **Emissions Control**: The most important section in environmental responsibility involves the measurements of the emissions. Emission control KPIs enables businesses to track their carbon footprint, adherence to environmental provisions and how to minimize on the emission.

  Carbon Footprint per Unit Produced: This KPI determines the number of carbon dioxide (CO2) emissions that have resulted per unit of product produced. It aids the companies to realize the effects they have on the environment and do what they can to minimize the emissions per unit of the product.

Emission Reductions Realized via Clean Technologies: This KPI will monitor the amount of reduction of emission due to the adoption of clean technologies, including carbon capture systems, integration of renewable energy and energy-efficient machines. Through this KPI, companies are able to gauge whether their sustainability efforts are working in cutting emissions.

International Emissions Standards: Firms should adhere to the local and international standards of emissions, including the Paris Agreement or the national carbon trading systems. This KPI follows the compliance of a company with the legal emissions threshold established by these regulations [14].

Example: An industrial plant may install AI-enabled carbon emissions monitoring systems that would monitor the carbon emissions in real time. Such systems give information on the cause of excess emissions; the plant can make remedial measures and enhance control of emissions.

6.6.2 Data Driven Decision Making

The introduction of IoT technology, systems with the focus on AI and automation of manufacturing and business processes enables the collection of data in real-time, which is the secret to tracking sustainability KPIs. Using data-driven decision-making would allow business to continuously monitor their sustainability indicators and correct them in time to improve their environmental performance.

- **Real time Data Monitoring**: In the Industry 5.0, the IoT sensor and AI data collection would allow the business to view the sustainability indicators, such as energy consumption, waste, and emissions, in real-time. This enables to make proactive decisions and intervene early enough to remove inefficiencies and ensure that the sustainability goals are met.

  Example: The use of IoT in a factory can be used to track real-time energy consumption, water and emissions. This data can be analyzed with the help of the AI algorithm and recommend the methods of waste reduction, energy use, and emissions.
- **Sustainability Optimization using Predictive Analytics**: The machine learning and AI models can also be utilized to predict the future sustainability performance based on the past data and trends. Predictive analytics enable the company to know how much energy it would use over the years, the amount of waste they would generate and the amount of emissions they would make and implement steps to ensure they remain on the right path towards their sustainability targets [15].

  Example: A retailer can use predictive analytics to forecast the energy usage of its supply chain operations and trace the potential inefficiency before it occurs. This will help the company to streamline its energy and logistics plans.

Figure 6.4 shows how sustainability is an essential part of Industry 5.0 that focuses on the use of Key Performance Indicators (KPIs) to ensure environmental improvement. It also emphasizes major KPIs like energy efficiency, waste minimization,

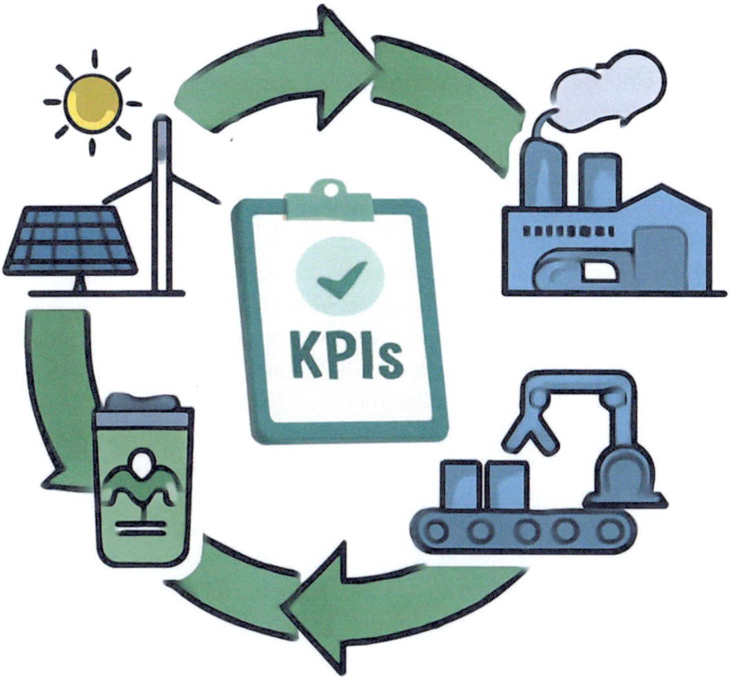

Fig. 6.4 Empowering sustainability with key performance indicators

and emission control. The imagery demonstrates such areas as solar panels, wind turbines, recycling, industrial processes, and automation, all of which are aimed at the realization of the sustainability objectives in the form of data-driven insights.

6.6.3 Advantages of Sustainability KPIs

There are a great number of benefits associated with KPI monitoring on sustainability to business such as improved performance, accountability and transparency. These benefits help companies to achieve their sustainability goals besides leaving a difference to the environment.

- **Performance Tracking**: Sustainability KPIs may provide actual and quantifiable information to the companies to assess their performance within the environment. Observation of such measures will help organizations to understand where they are performing well and where they need to improve. This also allows the businesses to make sound decisions, enhance strategies and stream processes to produce better environmental outcomes.

- **Improved Decision-Making**: Sustainability KPIs have the ability to provide a data-driven information, which can be used to guide businesses to make improved decisions with regard to the distribution of resources, energy use, waste, and emission reduction. Real-time tracking of performance enables the companies to simplify their operations and reduce on inefficiencies in order to make them have minimal effects on the environment.

  Example: A manufacturing company may use sustainability KPIs to monitor the energy consumptions and calculate the most active times of the consumption. Energy use optimization where the energy consumption happens at the peak will assist the company to reduce the expense of energy and increase sustainability.
- **Accountability and Transparency**: The Sustainability KPIs reporting will help the business demonstrate that they are environmentally friendly and it will also provide transparency to the stakeholders including the consumers, the regulators and the investors. The reporting is open, and this generates confidence and improves the association with the customers and investors interested in integrating sustainability in their purchase and investment patterns [16].

  Example: A company that publicly makes its environmental reduction target, waste recycling and the use of renewable energy resources, earns the trust of consumers and investors. This kind of transparency can also improve the company image and attract environmentally friendly customers and investors.

6.7 Life Cycle Assessments (LCA) and Cost–Benefit Analysis (CBA) of Sustainable Practices

In the quest to sustainability, the business will have to consider both the environmental and the economic consequences of their actions. Industry 5.0 embraces the use of sophisticated analysis tools in the establishment of the long-term implications of sustainable practices, including Life Cycle Assessment (LCA) and Cost–Benefit Analysis (CBA). The tools will provide the information that businesses need to make wise decisions in the sustainability initiatives to use and it will be founded on the balance between the economic viability and the environment. LCA and CBA enable business to know more about the trade-offs, explore on the areas to improve and plan their investments in sustainability.

6.7.1 Life Cycle Assessment (LCA)

Life Cycle Assessment (LCA) represents comprehensive process, which evaluates the environmental performance of a product or service by taking into consideration the entire lifecycle of the product or service i.e. extraction of the raw material, production, distribution, use and the disposal of the trace elements at the end of life. LCA helps the firms to understand the effects of their services on the environment and, therefore, demonstrates where sustainability can be made better and where the environmental performance is to be increased. The tool is gaining more and more applications in Industry 5.0 so that products and processes are developed in a sustainable way.

- **In-depth Environmental analysis**: LCA examines each phase of the lifecycle of a product to determine its effects on the environment, i.e. extraction, energy usage, waste products and emissions. This holistic approach would help businesses to recognize some of the most critical aspects, which could be refined to ensure that environmental degradation is minimized.

  Example: To assess the environmental impact of acquiring raw materials (e.g. mining of metals), production (e.g. energy consumption, waste produced), use (e.g. power consumed, etc.) and due to disposal or recycling of end products of electronic products, a firm involved in the production of electronic products can use LCA. Analyzing such stages, the company can find ways to use more sustainable materials, consume less energy, or make their products more recyclable.
- **Sustainable manufacturing and Product Design**: LCA promotes the use of cradle-to-cradle design in the business world where all the components of the lifecycle of a product are sustainable. This is especially relevant in Industry 5.0 since it incorporates the concept of circularity according to which products are not limited to their end useful life; they can be reused, remanufactured, or recycled.

  Product Design: LCA influences firms to design products that are environmentally friendly in terms of the materials they are made out of, the energy they consume in the manufacturing process as well as the chances of being recycled at the end of their lifecycle. Businesses can lower their carbon footprint, conserve resources and reduce waste, by designing products that are easy to repair, upgrade, or recycle [17].

  Example: In automotive industry, electric vehicle (EV) manufacturers can use LCA to measure the environmental effects of battery manufacturing, manufacturing energy and energy efficiency of the vehicle. The outcome of the LCA can be used to make decisions at the design level to maximize the lifecycle of the car like choosing materials that could be recycled more easily or making the energy storage system more efficient.

6.7.2 Cost Benefit Analysis (CBA)

Cost–Benefit Analysis (CBA) is an important economic instrument which guides the organization to analyze the sustainability project in terms of financial viability. CBA evaluates the cost of putting in place a sustainable practice against its long-term advantages, which include low operation costs, compliance with the regulations, and environmental benefits. Such analysis helps companies to make the data-based decisions related to green technologies and sustainable practices to be most economically feasible and aligned to the objectives of the company.

- **Economic feasibility of Green Projects**: Although sustainable practices are usually associated with an initial investment, e.g., the expenses of new technologies, machinery, or infrastructure, CBA allows businesses to estimate whether the long-term payoff, e.g., lower operating expenses, energy savings, or fines evasion, is worth the initial investment.

  Example: A firm that is contemplating the use of solar panels in its factory may apply the concept of CBA to compare the cost of purchasing the solar panels with the amount of money that will be saved in paying the electricity bills after several years. Also, the company can look at the possible tax breaks or rebates on the adoption of renewable energy, and the long-term benefit of reducing its carbon footprint.
- **Long-term Benefits Evaluation**: CBA also allows the companies to realize the wider long term economic benefits of the sustainable practices. Some of the benefits are low waste management expenses, energy savings, better regulatory compliance, and better brand value. Companies can also enhance their operations by going green or enhancing their sustainability, and in the process develop competitive advantage in the market [18].

  Example: CBA can help a manufacturing plant to determine the economic implication of using waste-to-energy system. Although the cost of investing in technology might be high, the long-term returns, which include the reduction of the expense of disposing of the trash, saving on energy expenses, and renewable energy production, can justify the investment.

Figure 6.5 visually conveys the integration of sustainability evaluation tools in Industry 5.0, emphasizing the role of Life Cycle Assessment (LCA) and Cost–Benefit Analysis (CBA). On the left, a pie chart and bar graph represent data analytics used in evaluating environmental and economic impacts. In the center, a globe icon symbolizes the environmental perspective of sustainability. On the right, a scale balances a plant (representing environmental sustainability) with a money bag (representing financial considerations), illustrating the cost–benefit analysis of green technologies and sustainable practices while considering the industry 5.0.

6.7.3 Industry 5.0 Advantages of LCA and CBA

LCA and CBA are both vital in assisting businesses to make sound judgments concerning the sustainability activities. The advantages of these tools are quite enormous as they include better decision-making, cost reduction and better competitive edge.

- **Informed Decision-Making**: LCA and CBA inform businesses on information that they require to make strategic decisions on investments in sustainability. The evaluation of environmental and economic effects of the possible initiatives allows the companies to find the most economic and environmentally friendly solutions.

Fig. 6.5 Harnessing LCA and CBA to drive sustainable innovations in Industry 5.0

Example: A company may realize, through an LCA, that it can use materials that are locally sourced (which will reduce its emissions during transportation) compared to importing raw materials produced abroad, which may appear to be cheaper on the surface. In the same way, CBA may be used to determine whether investment in renewable energy systems will lead to long-term savings or other energy-saving projects should be given priority.

- **Cost Savings**: The result of sustainability is usually low operational costs in the long run. Companies can reduce their operating costs by adopting energy-efficient technologies, minimizing wastes and maximizing resources. Such savings will be able to cover initial investments and provide financial returns in the long term.

  Example: An energy-efficient machinery that is adopted by a company will enable the company to save in terms of energy consumption and electricity bills thus resulting in a huge saving over the lifetime of the machinery. Such savings have the potential of making this investment in new and sustainable technologies profitable in the long term.
- **Competitive Advantage**: When a company is able to show the economic and environmental value of their sustainability efforts then they have a great competitive edge. With consumers and investors becoming more interested in sustainability, businesses that are pioneers in green operations get environmentally friendly customers and investors, which improves their market share.

  Example: Companies that adhere to the sustainability norms, reduce their carbon footprint and publicly announce their achievements are more reputable and have more opportunities to build loyalty among environmentally responsible consumers. This can lead to an increased market share and brand equity.

- **Regulatory Compliance and Risk Management**: By incorporating sustainability in the operations of the business, companies can comply with the increasing number of environmental rules and regulations. The sustainability programs also allow firms to escape the chance of being fined or punished later on when they take action in advance and deter environmental degradation [19].

  Example: A company that has made investments to reduce its emission to comply with the local or international environmental standards will not pay huge fines or penalties and will even have the right to continue operating.

6.8 Learning Outcomes

After going through this chapter, the reader will be in a position to:

- Determine and elucidate the fundamental sustainability principles that are propelling Industry 5.0 such as energy efficiency, waste reduction and controlling emissions.
- Learn the significance of the circular economy and its integration into the contemporary manufacturing cycle to lessen the use of resources and impact on the environment.
- Consider the advantages and difficulties of adopting sustainable practices within the industrial systems and especially regarding efficiency of operation, regulatory compliance and consumer confidence.
- Examine how advanced technologies, including IoT, AI, and renewable energy, contribute to sustainability objectives in Industry 5.0.
- Get familiar with the regulatory environment and know how businesses can meet national and international environmental standards and conventions.
- Use Key Performance Indicator (KPIs) and data-driven intelligence to monitor the sustainability performance and enhance the way operations take place.
- Employ Life Cycle Assessment (LCA) and Cost–Benefit Analysis (CBA) as an instrument of measuring the environmental and economic effects of sustainable initiatives.

6.9 Review Questions

1. What does the industry 5.0 redefine regarding sustainability in the manufacturing process when compared to the past industrial revolutions?
2. What are the main advantages and issues of applying the concept of circular economy in industrial manufacturing, including reuse, remanufacturing, and recycling?
3. (a) Briefly describe how advanced technologies like IoT, AI and renewable energy can be used to enhance energy efficiency and emission control in Industry 5.0.

(b) What are the ways through which these technologies can assist industries to cut their dependence on fossil fuels and support the global sustainability objectives?
4. What are the cost and regulatory compliance and reputation advantages of sustainability practices of Industry 5.0 to companies?
5. (a) What are the effects of international agreements, like the Paris Agreement, on the industrial sustainability strategies, and
   (b) How do national environmental regulations contribute to the development of the approach to emission control and resource management in Industry 5.0?
6. Briefly describe how real time data and IoT devices can assist businesses to track energy consumption, waste management and emission control.
7. (a) What are the ways of applying Life Cycle Assessment (LCA) and Cost–Benefit Analysis (CBA) in Industry 5.0 to measure the long-term effects of sustainability measures?
   (b) Please give an example of a business decision in which LCA and CBA may be used to inform the implementation of a sustainable practice.
8. What is the role of the principles of a circular economy implemented with the help of advanced technologies in decreasing the dependence on virgin materials and minimizing environmental pollution?
9. How can businesses take a competitive advantage of sustainability practices in Industry 5.0?
10. (a) What are the possible barriers that industries can encounter on the way to a circular economy model in Industry 5.0?
    (b) Write about the technological, logistical and financial barriers and how these barriers can be addressed.

Appendix: Solution Manual

1. **What does the industry 5.0 redefine regarding sustainability in the manufacturing process when compared to the past industrial revolutions?**
   Solution: Industry 5.0 redefines sustainability by integrating advanced technologies like IoT, AI, and renewable energy to optimize energy usage, minimize waste, and reduce emissions. Unlike previous revolutions, sustainability is a core driver of innovation, not an afterthought. Businesses now focus on smart energy management systems, renewable energy integration, and the use of AI to prevent waste and control emissions. These practices are aimed at minimizing the environmental footprint while maintaining operational efficiency.
2. **What are the main advantages and issues of applying the concept of circular economy in industrial manufacturing, including reuse, remanufacturing, and recycling?**

Solution: Key benefits of the circular economy include reduced resource consumption, sustainable product design, and lower environmental impact. Challenges include the need for significant infrastructure changes, technological investment, and collaboration across supply chains. Industries such as automotive and electronics have successfully implemented circular models by remanufacturing components and integrating recycling processes, extending product life cycles and reducing waste.

3. **(a) Briefly describe how advanced technologies like IoT, AI and renewable energy can be used to enhance energy efficiency and emission control in Industry 5.0.**

   Solution: For example, AI-powered energy management systems optimize factory operations, ensuring minimal energy usage while maximizing output.

   (b) What are the ways through which these technologies can assist industries to cut their dependence on fossil fuels and support the global sustainability objectives?

   Solution: Renewable energy integration, such as solar and wind power, reduces reliance on fossil fuels, helping industries reduce their carbon footprint and support global sustainability efforts.

4. **What are the cost and regulatory compliance and reputation advantages of sustainability practices of Industry 5.0 to companies?**

   Solution: Sustainability practices benefit companies by reducing operational costs (e.g., energy savings and waste reduction), ensuring compliance with environmental regulations, and enhancing their brand reputation. Customers are increasingly prioritizing sustainability, and businesses that align with these values enjoy increased loyalty, enhanced public perception, and a competitive edge in the marketplace.

5. **(a) What are the effects of international agreements, like the Paris Agreement, on the industrial sustainability strategies?**

   Solution: Global agreements like the Paris Agreement set targets for reducing emissions, influencing businesses to adopt sustainable practices.

   (b) How do national environmental regulations contribute to the development of the approach to emission control and resource management in Industry 5.0?

   Solution: National standards help companies comply with local regulations regarding waste, energy use, and emissions. Industry 5.0 technologies such as carbon capture systems and renewable energy integration assist businesses in meeting these global and national sustainability targets.

6. **Briefly describe how real time data and IoT devices can assist businesses to track energy consumption, waste management and emission control.**

   Solution: KPIs for sustainability include energy consumption per unit of production, reduction in carbon emissions, percentage of renewable energy used, and waste recycled. IoT devices track real-time data on energy use, and AI algorithms analyze this data to identify inefficiencies and suggest improvements. This allows businesses to monitor progress toward sustainability goals and make data-driven decisions.

7. **(a) What are the ways of applying Life Cycle Assessment (LCA) and Cost–Benefit Analysis (CBA) in Industry 5.0 to measure the long-term effects of sustainability measures?**

   Solution: LCA evaluates the environmental impact of a product throughout its lifecycle, while CBA compares the costs and benefits of implementing sustainability initiatives.

   (b) Please give an example of a business decision in which LCA and CBA may be used to inform the implementation of a sustainable practice.

   Solution: For example, a company may use LCA to assess the environmental impact of a new product design, considering raw material extraction, energy use, and disposal. CBA helps determine if the long-term savings from sustainability initiatives outweigh initial costs, guiding decision-making.

8. **What is the role of the principles of a circular economy implemented with the help of advanced technologies in decreasing the dependence on virgin materials and minimizing environmental pollution?**

   Solution: Traditional manufacturing relies heavily on extracting virgin materials and energy-intensive processes, contributing to environmental pollution. Industry 5.0's approach emphasizes circular economy principles, where materials are reused, remanufactured, and recycled, reducing the need for raw materials and minimizing pollution. Technologies like AI, IoT, and recycling automation enable this shift, reducing waste and conserving resources.

9. **How can businesses take a competitive advantage of sustainability practices in Industry 5.0?**

   Solution: Businesses that prioritize sustainability can differentiate themselves in the market, meeting the growing consumer demand for eco-friendly products. Sustainability practices can lead to increased market share, customer loyalty, and positive brand perception, ultimately providing a competitive advantage in both local and global markets.

10. **(a) What are the possible barriers that industries can encounter on the way to a circular economy model in Industry 5.0?**

    Solution: Transitioning to a circular economy model presents challenges such as high initial investment, the need for specialized technologies, and the coordination of supply chains.

    (b) Write about the technological, logistical and financial barriers and how these barriers can be addressed.

    Solution: Overcoming these challenges requires collaboration with suppliers, investment in research and development, and a commitment to long-term sustainability goals. Financial incentives, such as government subsidies for green technologies, can also help ease the transition.

Bibliography

1. Rame R, Purwanto P, Sudarno S (2024) Industry 5.0 and sustainability: an overview of emerging trends and challenges for a green future. Innov Green Develop 3(4):100173
2. Van Erp T et al (2024) Industry 5.0: a new strategy framework for sustainability management and beyond. J Clean Prod 461:142271
3. Szara K (2025) Resilience in Industry 5.0—systematic literature review. Scientific Papers of Silesian University of Technology. Organization & Management/Zeszyty Naukowe Politechniki Slaskiej. Seria Organizacji i Zarzadzanie 224
4. Gaikwad S, Anerao K (2025) Industry 5.0 Innovations for waste management and circular economy in sustainable food systems: a systematic review. In: BIO web of conferences, vol 178. EDP Sciences
5. Hammad M et al (2025) Industry 4.0 in pharmaceuticals: a new era of smart manufacturing. In: Smart manufacturing blueprint: navigating Industry 4.0 across diverse sectors. Springer, Cham, pp 165–211
6. Hsu C-H et al (2024) Enabling Industry 5.0-driven circular economy transformation: a strategic roadmap. Sustainability 16(22):9954
7. Khan WA et al (2025) Remanufacturing. In: Computer aided engineering design and manufacturing: a fourth industrial revolution perspective. Springer, Cham, pp 187–211
8. Hammad M et al (2023) Framework for the implementation of smart manufacturing systems: a case in point. Processes 11(5):1436
9. Noman AA et al (2022) Machine learning and artificial intelligence in circular economy: a bibliometric analysis and systematic literature review. Ann Emerging Technol Comput (AETiC) 6(2):13–40
10. Shafique I, Kalyar MN, Mehwish N (2021) Organizational ambidexterity, green entrepreneurial orientation, and environmental performance in SMEs context: examining the moderating role of perceived CSR. Corp Soc Respons Environ Manage 28(1):446–456
11. Lopez-Merino P, Charlier C, Guerci E (2025) State-of-the-art review on circular economy models for plastic waste
12. Tessitore S, Corsini F, Iraldo F (2023) Exploring the alignment and misalignment in the transition to a circular economy within private companies and public organizations. J Clean Prod 430:139528
13. Maraseni T, Reardon-Smith K (2019) Meeting national emissions reduction obligations: a case study of Australia. Energies 12(3):438
14. Movahed AB, Nozari H, Bakhshi Movahed A (2025) Systematic review of smart factories production in Industry 5.0. arXiv preprint arXiv:2504.21525
15. Hammad M et al (2025) Textiles and apparel: adapting smart manufacturing for creativity and customization. In: Smart manufacturing blueprint: navigating Industry 4.0 across diverse sectors. Springer, Cham, pp 245–288
16. Czvetkó T, Sebestyén V, Abonyi J (2025) Key factors of Industry 5.0-based organizational sustainability. Technol Soc 102966
17. Hossain MdU, Thomas Ng S (2019) Influence of waste materials on buildings' life cycle environmental impacts: adopting resource recovery principle. Resour Conserv Recycling 142:10–23
18. Singh S, Yadav R, Narain Singh A (2025) Circular practices in the agricultural sector: current challenges and opportunities. In: Circular economy and green transition in the global south: perspectives and prospects, pp 169–197
19. Khan WA et al (2025) Industry 5.0. In: Computer aided engineering design and manufacturing: a fourth industrial revolution perspective. Springer, Cham, pp 523–535

Part V
Data, Cybersecurity, and Sector Applications

Chapter 7
Data Analytics, Cybersecurity, and Privacy

7.1 Objectives

- To specify and discuss the role of data analytics in Industry 5.0, with a focus on the shift to prescriptive analytics.
- To get to know the different cybersecurity threats posed by the interconnectedness of human–machine systems in Industry 5.0.
- To investigate the privacy issues of personal data collection and usage in smart environments and the way in which these issues may be handled ethically and legally.
- To define and present the most appropriate practices and technological protection measures in the field of cybersecurity and data privacy in Industry 5.0.
- To examine risk assessment models and breach cost and how companies can determine and reduce the risk of cybersecurity.
- To offer practical recommendations that companies can follow to achieve proper data analytics, cybersecurity, and privacy protection strategies and comply with the regulatory requirements.

7.2 Scope

Since Industry 5.0 is increasingly becoming compatible with advanced technologies like artificial intelligence (AI), the Internet of Things (IoT), and cyber-physical systems, it also presents a new set of challenges and opportunities in the context of data analytics, cybersecurity, and privacy protection. The mass of information produced by networked devices and the growing complexity of online threats have made strong data management policies and security measures more relevant than ever. It is the task of organizations to overcome these difficulties to guarantee the

M. Hammad et al., *Industry 5.0: The Human-Centric Future of Industry*, Springer Series in Advanced Manufacturing, https://doi.org/10.1007/978-3-032-12622-1_7

security, integrity, and privacy of their systems and take the best of the opportunities advanced analytics and interconnected technologies provide. Here, we discuss the changing role of data analytics, cybersecurity risks of human–machine systems, privacy issues of Industry 5.0, as well as the measures that can assist in reducing the risks.

7.3 The Role of Data Analytics: The Future of Descriptive to Prescriptive Analytics of Industry 5.0

Data analytics has taken the position as an important part of contemporary business decision-making. Since Industry 5.0 is about the cooperation of people and machines, data analytics is much more than just simple reporting and other forms of basic descriptive analytics. Within the Industry 5.0 framework, data analytics is an essential aspect as it helps to develop operational strategies, enhance the performance of the businesses, and forecast future tendencies. High-tech tools such as AI and machine learning enable businesses to draw more practical conclusions based on the tons of data produced by the connected systems. The role of analytics can be divided into three broad phases, namely, descriptive, predictive and prescriptive analytics [1].

i. **Descriptive Analytics**

Descriptive analytics is the basis of data analysis in businesses. It is a traditional method which aims at summarizing the data of the past and identifying trends, patterns and knowledge that indicate how the past was performed. The questions that are answered by descriptive analytics include the following: What happened? and What is presently the case? It is the initial phase of data-driven decision-making because it gives organizations an in-depth insight into their operations, customers, and performance indicators.

Example: A firm in the manufacturing sector can utilize descriptive analytics to study historical production quantities, the fault level, energy use, or costs incurred in the process. Such insights provide an overview of the performance of the business within a given timeframe and point out where the business needs to make some improvements. As an example, a factory can discover that the volume of production decreased during some months or that energy consumption increases during a particular time of the day, which can be used as a starting point to analyze the situation.

Descriptive analytics is useful in giving some insights, but it fails to offer anything above summarizing past events. It is not able to tell what will happen in the future or what can be done on the basis of such information. Therefore, organizations that would like to enhance their decision-making capacities need to shift the current descriptive analytics to the next level, predictive and prescriptive analytics.

ii. **Predictive Analytics**

Predictive analytics based on AI, machine learning and big data analytics is an important part of the journey that organizations are going through in Industry 5.0. Predictive analytics uses past data, statistical models, and machine learning algorithms to predict future trends possible problems/risks. Predictive analytics can assist organizations in looking ahead and preparing to counter the challenges that may come in future by studying the trends in data that has already been collected.

Case in Point: Predictive maintenance models are common in manufacturing, where they are used to ensure that maintenance activities are not carried out on an ad-hoc basis, which is costly. Machine learning systems may use historical data stored in IoT sensors installed in machines to make predictions about machine failure probabilities, given usage patterns, vibration data, temperature variations and other performance indicators. This enables timely maintenance and repair, thus minimizing the interference with production and the ease of the manufacturing process. They can also be applied to predict future product demand, how much time an inventory has left on the shelf or how many staff members will be required in the future based on the past.

It is priceless to be able to foresee possible problems before they actually turn out, and it is particularly important in those industries where a delay or a stop can result in a significant loss of finances. Predictive analytics gives organizations the vision they require to deal with issues before they happen, and this not only increases the efficiency of operations but also reduces the chances of unexpected failures.

iii. **Prescriptive Analytics**

Prescriptive analytics is one step ahead of predictive analytics, as it does not only predict the future but also makes suggestions as to what actions are to be taken in order to maximize the results. The question that prescriptive analytics answers is what should we do regarding it? Or What is the best thing to do? Prescriptive analytics can assist organizations in making data-driven decisions that result in better performance and efficiency by integrating advanced algorithms, machine learning models, and real-time data.

Example: In a smart factory, prescriptive analytics could be applied to make real-time adjustments to the production schedules using up-to-date information about the performance of machines, the levels of inventory and the productivity of the workers. As an example, prescriptive analytics can recommend maintenance activities or recommend redistribution of resources to prevent delays in case AI-powered systems determine that a machine is not working as efficiently as it should. Likewise, it will be able to optimize the supply chains by suggesting changes in inventory levels depending on the variable demand and production rates. Prescriptive analytics assists organizations in determining the best course of action to take in order to enhance productivity, cut down on costs and meet business objectives, considering several variables and constraints.

Prescriptive analytics would be very handy in cases where you have an active situation whereby the decisions made must consider a large number of factors and interdependencies. It provides a pragmatic expertise that can assist organizations to

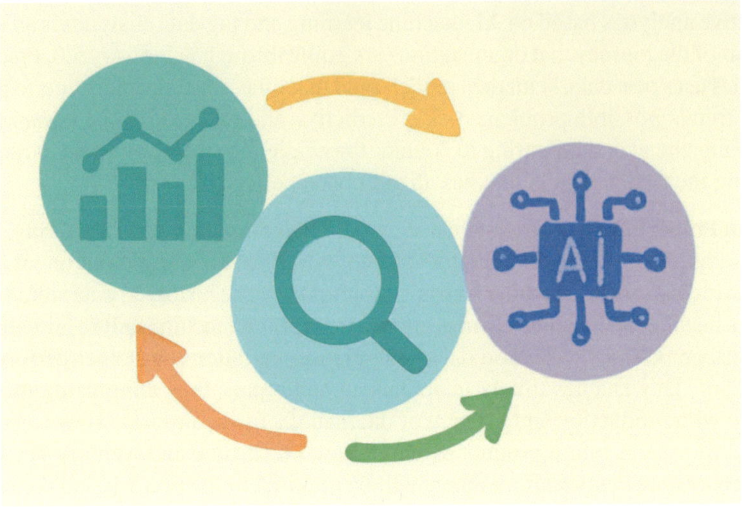

Fig. 7.1 Harnessing data analytics for smarter decision-making in industry 5.0

streamline their operations, allocate their resources and provide improved outputs in an immense number of business procedures.

Figure 7.1 depicts the progress of the data analytics in Industry 5.0, the transformation of Descriptive Analytics summarizing the past, into the Predictive Analytics, forecasting the future tendencies, and the last one, Prescriptive, the recommendation of what is best to do to optimize operations. The arrows and the icons are interconnected to highlight the cyclic and progressive aspects of analytics in the contemporary industries in a bid to enhance decision-making and operational efficiency.

7.3.1 Benefits of Data Analytics for Industry 5.0

There are numerous benefits that incorporating improved data analytics in Industry 5.0 has to organizations, such as improved operation efficiency and enhanced decision-making. As the trend in utilization of analytics by businesses changes into prescriptive analytics, businesses will be in a position to maximize the potential of their data and gain a strategic benefit.

- **Increased Decision Making**: Organizations can make informed decisions that are better with the application of the advanced analytics. Descriptive analytics provide the background information, but predictive and prescriptive analytics provide the possibility to predict the future trend and suggest the most appropriate course of actions. This generalized decision making process ensures that organizations are ready to deal with issues before they occur and apply the business strategies in the most desirable manner.

- **Operational Efficiency**: Predictive and prescriptive analytics can help identify the bottlenecks, eradicate waste, and utilize the resources more efficiently across the operations. Another illustration is in the manufacturing sector where predictive analytics will be used in forecasting when the equipment would break and prescriptive analytics could be deployed to modify the production lines to realise a better throughput and fewer downtime. This leads to an improved utilization of the resources, low operation costs and quality of the product.
- **Competitive Advantage**: The companies, which use the power of sophisticated analytics have the chance to monitor the appearance of new tendencies, determine new chances, and make reasonable decisions faster than other companies. By having the benefit of anticipating consumer needs or a certain market, organizations can react very quickly to a change in the market, or even a need of a customer, thus providing the organization with a competitive edge [2].

7.4 Risks of Cybersecurity: Weaknesses of Interconnected, Human–Machine Systems

Industry 5.0 is another stage of interconnectivity when modern technologies such as artificial intelligence (AI), the Internet of Things (IoT), robotics, and cyber-physical systems will collaborate closely with human employees. Although these innovations are of immense benefit, they pose many cybersecurity risks. The interdependence of devices, machines, and systems causes an increased attack surface, and the industries are more exposed to various cyber threats, such as unauthorized access, data breaches, and other attacks that might lead to the loss of digital assets and physical infrastructure. With the rise in complexity of human–machine systems, it is important to make sure that such interconnected spaces are secure to avoid interruption, loss of intellectual property and compromise of human safety.

7.4.1 Weaknesses of Human Machine Systems

The perfect marriage between man and machine in Industry 5.0 leads to interrelated systems that should be handled with great consideration for security. Nevertheless, there are a number of weaknesses which can be used by cybercriminals. Such vulnerabilities include IoT device weaknesses, as well as the human factor, which is one of the most popular risk factors in cybersecurity.

- **IoT Devices**: The Internet of Things (IoT) has been used in Industry 5.0 as the basis of smart systems which capture and relay real-time data. The operations of such industries as manufacturing, healthcare, and agriculture include IoT devices, sensors, smart machines, and wearable devices. These devices impart real-time information and management and help in simplifying the processes and making them efficient. However, they are equally vulnerable to cyber criminals.

Lack of Security: IoT devices are usually built to be functional and cost-effective at the cost of sound security. Such devices can be poorly encrypted, lack a strong authentication method, or their software is outdated, which makes them prone to exploitation. After being compromised, IoT devices can be used as entry points by attackers to penetrate the greater network and perform even more serious attacks, including data theft or ransomware.

Massive Attack Surface: The massive amount of IoT devices in a typical Industry 5.0 setup will make it more probable that there will be some vulnerability. Any of the connected devices is a potential target of cyberattacks, and in case one of the devices is hacked, it can be used as a steppingstone to hack other connected systems, increasing the risks even more. As an illustration, breaking into a wearable device of a factory employee may allow the attacker to obtain access to the whole industrial network.

- **Industrial Control Systems (ICS)**: ICS plays a very important role in industries like manufacturing, energy, and utilities. Such systems manage critical infrastructure such as machinery, power grids and production lines. ICSs are, however, regularly linked to company networks and the internet in general to be monitored and controlled, and this makes them prone to cyber threats.

  Legacy Systems: A lot of industrial control systems are using legacy technology which was not cybersecurity-driven. Such obsolete systems might not have the latest security measures, and thus, it is easy to attack them by cybercriminals. In certain instances, manufacturers have no plans of abandoning these systems since the process of upgrading to safer platforms is expensive and complicated.

  Absence of Security Integration: ICS are usually isolated to enterprise networks and, in that regard, remain unintegrated with contemporary cybersecurity tools. Such systems might not be as well-protected against security attacks as IT networks, so that they can become victims of cyberattacks, including ransomware data, manipulation and even sabotage. A hacker who has access to an ICS may alter the production processes, malfunction the equipment, or even interrupt the entire supply chain [3].

- **Human Error**: Nevertheless, even in the era of Industry 5.0, there is a major weak point in human workers. Automation and AI are becoming more capable of performing repetitive tasks, but in most spheres, human supervision and judgment are of utter importance. Cybersecurity risks may be caused by human error, like when systems are misconfigured, when people are negligent, or when there is poor cybersecurity training.

  Misconfigured Systems: The human workers can make mistakes in the security settings or leave some important systems unsecured. An example is that a worker can accidentally turn off firewalls, leave sensitive information on the open internet, or leave software vulnerabilities without patching them, which allows cybercriminals to attack the system.

Ineffective Password Management: Ineffective password management is still one of the most widespread weaknesses in most organizations. Even though there are more secure forms of authentication, such as multi-factor authentication (MFA), employees can continue to use easy or the same passwords on several systems, and attackers can more easily gain unauthorized access.

Lack of Training: Most of the cybersecurity breaches occur because of human employees being lured by phishing, social engineering or other types of attacks. Unless employees are trained in the best way to identify and act against such threats, they may end up giving cybercriminals an opportunity to gain access to networks without knowing.

7.4.2 Categories of Cybersecurity Risks

The interdependence of human and machine systems creates a diversity of cybersecurity threats that may create serious disturbances, loss of revenue, and reputations. Ransomware attacks, data breaches, and denial-of-service (DoS) attacks are the most common forms of cyber threats in the industry 5.0 setting.

- **Ransomware Attacks**: Ransomware attacks entail malicious software that locks data in an organization and requires payment in order to unlock it. Such an attack in an interconnected, human–machine system can bring disruption on a large scale, and it may paralyze entire industrial operations.

  Cascading Effects: A ransomware attack on one machine or other device in a highly interconnected system can be easily transmitted throughout the network to other machines, production lines or facilities. It may lead to closure of vital processes, stalled productions and huge monetary losses. To take one example, an attack on a manufacturing plant by a ransomware program will lock up automated machines so workers cannot access critical systems or data required to operate.

  Downtime: Downtime that will be caused by the attack of ransomware can be expensive. In the event that industrial control systems or data that is critical to business operations is encrypted, the organizations may be disrupted for weeks as they recover systems or negotiate with cybercriminals. Also, the long-term consequence of reputational damage may be long-lasting damage to customer trust and loyalty.
- **Data Breaches**: A great quantity of sensitive information is constantly exchanged among human employees, machines, and cloud services in an environment of Industry 5.0. This information may consist of intellectual property, trade secrets, personal customer data or business proprietary strategies. The consequences of cybercriminals accessing such information can be disastrous to companies.

Intellectual Property Theft: Intellectual property theft can take place where innovation and proprietary designs play a vital role in competitive advantage, as in the case of manufacturing, and data breaches may result in the theft of trade secrets, designs or product blueprints. This information may be sold to the competition, and this will cause it to lose market position and also reduce profitability [4].

Exposure of Personal Information: By exposing the data, personal information may also be revealed in some industries, such as healthcare, where the data of patients is shared and stored electronically. Not only will this be against privacy regulations such as GDPR or HIPAA, but it also may lead to the irreversible destruction of the people involved and the reputation of the organization.

- **Denial-of-Service (DoS) Attacks**: Denial-of-service (DoS) attacks are performed to flood a system with too much traffic and make it unusable to legitimate users. Such attacks may be used in Industry 5.0 and be directed to vital manufacturing systems, supply chain management systems, or energy grids.

  Operational Disruption: A DoS attack on one of the key industrial systems can render it unusable, which will lead to extreme disruption in manufacturing or supply chain operations. To illustrate, in case a DoS attack is directed against the automation system of a smart factory, it may stop production lines, cause delays in orders, and trigger failure to meet deadlines and financial losses.

  Financial Losses: In addition to the direct effect of the disrupted operations, organizations which became targets of DoS attacks can face extra expenses on recovery, such as repairing the system, taking legal measures, and compensating the customers. The more the attack drags on, the more the cost is imposed on the organization [5].

This visual 7.2 illustrates the most important cybersecurity threats involved in the interconnected systems of the Industry 5.0 with a focus on the possible vulnerabilities created by IoT devices, industrial control systems, and human factors. The picture relies on different icons to represent the cyber dangers such as unauthorized access, ransomware infections, data breaches, and denial-of-service (DoS) attacks capable of affecting both data and physical resources (Fig. 7.2).

7.4.3 Advantages of Cyber Security

Given the cybersecurity issue with the Industry 5.0, cybersecurity plays a vital role in ensuring the safety of organizational assets, business continuity, and capacity to comply with regulatory standards. The benefits of using strong cybersecurity systems in an environment that is connected between human and machine are the following:

- **Avoidance of Downtime**: A good cybersecurity measure will prevent interference associated with cyberattacks. The critical systems are secured against

Fig. 7.2 Key risks in industry 5.0's interconnected systems: Ransomware, data breaches, and human error threats

ransomware, data breaches and DoS attacks, which will allow businesses to have a smooth manufacturing process and other industrial activities. This reduces confinement of the operations and eases productivity, ultimately costs are saved and makes it more efficient.
- **Intellectual Property Protection**: In most firms, the intellectual property (IP) may be the most valuable asset, especially in the manufacturing sector, pharmaceutical sector and the technology sector. Good cyber security helps in deterring theft or unlawful access of proprietary information, designs, formula and trade secrets, hence ensuring that companies remain in a position to maintain the competitive edge and avoid losing revenues.
- **Customer Trust**: Customers are increasingly becoming insistent that organizations must keep their personal data safe and ensure that they have safe systems. Development of customer trust, particularly in healthcare, financial, and e-commerce industries, will result in a good cybersecurity position. This will assist organizations to create their image and enhance customer loyalty by demonstrating a good interest in securing sensitive information [6].

7.5 Privacy Concerns: Ethically, and Legally Dealing with Personal Data

As the world of Industry 5.0 actively moves towards the use of such innovative technologies as artificial intelligence (AI), Internet of Things (IoT), cyber-physical systems, and others, the collection, processing, and analysis of personal information became the key to creating intelligent, human-oriented environments. Even though there may be value delivery through such innovations with personalized services, optimized operations, and also improved user experiences, privacy matters are very serious when using personal data, particularly in smart environments. As the trend of gathering personal information becomes appealable using the various devices, sensors, and systems, it is significant that organizations implement ethical and legal frameworks to the protection of the rights of the people with regards to their privacy as well as ensuring that the organization is not contravening the laws on data protection. This section looks at the ethical management of personal information, legal concerns that companies ought to take note of and the benefits of taking the initiative against the concerns.

7.5.1 Ethical Treatment of Personal Data

With the growth of IoT-based devices, wearables and AI-based systems, organizations of the Industry 5.0 are collecting massive volume of personal data. This data is often associated with the actions of an individual, his/her health, location, or performance with devastating inquiries of surveillance, consent, and abuse. This personal data has to be handled in an ethical manner so as to foster trust and safeguard of individual rights.

- **Spying and Intrusion of Privacy**: Spying and Intrusion of Privacy: In this case, sensors can track employees in terms of their movement, health or work, and be provided in a smart factory. These systems are supposed to support the most effective processes and enhance safety and effectiveness. Nevertheless, there are issues of surveillance that may arise due to this continuous supervision, as it may violate the privacy of an employee or his or her sense of freedom. On the same note, wearable devices are used in medical care to gather information about health to personalize care or to monitor chronic illnesses, but the information may be sensitive and ethically needs to be handled with great intensity.

  Example: In a smart factory, wearable sensors could monitor the heart rate, body temperature or activity of a worker to make sure that he/she is working safely. Although this information is useful in enhancing productivity and detecting health hazards, it has been accompanied by a question concerning the autonomy of the employees and the level of monitoring. There is a thin line between safety and optimization and using this data and violating the rights of privacy by collecting too much or intrusive data.

  To respond to these issues, organizations are to introduce ethical rules for data collection that will not violate privacy and will not demean humans. Such guidelines are meant to make sure that the personal data is collected in legitimate business interest and that an individual is fully aware of the nature of data that is collected, how the data will be used and who will use it.

- **Anonymization and Minimization of Data**: The handling of data should be ethical and should consider reducing the volume of personal data that is collected and anonymizing the same as much as possible. Anonymization of personal information allows companies to minimize the threat of misuse or unauthorized access and, at the same time, extract useful information that can be used to improve optimization. The personal information is also not over-collected through data minimization, where only data that is necessary for a specific purpose is collected, which decreases the likelihood of personal harm in the case of data breach.

  Example: A firm might want to gather information about the activity or performance of workers rather than gathering information about their specific health conditions or location data, which might be useful to the company without infringing on the privacy of the worker. This may be achieved by combining data from bigger numbers of employees to study the trends without compromising the identity of the individual employees.
- **Information and Transparency**: Informed consent is one of the major principles related to the ethical management of personal data. People should be well aware of the information being gathered and the reasons behind the gathering, the intended use of the information and their rights concerning the information. Companies are advised to make the processes of consent transparent and easy to use so that people can easily choose whether to participate in data collection or not without being punished [7].

  Example: An intelligent device worn in a medical facility must have a privacy policy that is easily comprehensible and leave an opt-out option to the specific data gathering functions, i.e. tracking their sleep patterns or their physical activity, in case the user does not wish to take part in those parts of the monitoring.

7.5.2 Legal Considerations

In addition to the ethical aspect, Industry 5.0 also requires businesses to deal with complicated legal structures that govern the process of collecting, utilizing, and securing personal information. There exist different laws to protect the rights of consumers, and companies are required to adhere to these laws to ensure that they do not face the wrath of the law and gain the trust of their consumers.

- **General Data Protection Regulation (GDPR)**: General Data Protection Regulation (GDPR) is a very broad and universal data privacy policy in the world. It is applicable to every organization processing the personal data of the people in the European Union (EU), irrespective of the origin of the organization. GDPR has very demanding conditions for collecting, storing, processing, and sharing personal data.

  Important Provisions

Transparency: Companies should brief the individuals on the kind of data that they will be collecting and the uses that they will have.

Consent: People should give their explicit consent so that their data can be processed when sensitive information is involved (e.g., health information).

Data Minimization: Personal data should be collected only in the least amount required for a particular purpose.

Right to Access and Erasure: It gives people the right to access their data and demand its deletion anytime.

GDPR is a requirement for companies that have operations in the EU or with the EU. Non-compliance with such rules may lead to huge fines and tarnished reputation.

California Consumer Privacy Act (CCPA): The California Consumer Privacy Act (CCPA) offers similar protection to the residents of California in the United States. The CCPA entails the right of consumers to know about the collected personal data, the right of access and deletion of their data, and the right to say no to the selling of personal data. Although the CCPA is applicable only to businesses that are based in California, its impact is global since numerous organizations across the globe deal with California residents or have their operations within the state.

- **Data Ownership**: The other important legal aspect is data ownership. With organizations increasingly using digital twins and IoT devices as well as AI systems to gather data, the issue of data ownership becomes more complicated. In most instances, people might not be in a position to know the ownership rights of their data and how they can be utilized.

  Who Owns the Data? A vital question is the ownership of the data, particularly the personal one. Does the data belong to the person who created it (e.g. an employee with a sensor or a health device in a patient)? Or is it owned by the organization which collects the data (e.g. the employer or the healthcare provider)? Well-articulated policies should be put in place where the ownership of the data is clearly spelt out, and the ownership of the personal data is left to the discretion of the individuals.

  Third-Party Data Sharing: Third-party data sharing is the fact that, in many cases, IoT devices or AI systems share their gathered data with third-party vendors, including a cloud provider or analytics firm. This brings more legal issues in terms of control and safety of the data after it has been transferred out of the original organization. Policies on third-party data sharing and accountability should be clear to make sure that personal data is safeguarded all along the lifecycle [8].

Figure 7.3 illustrates the importance of handling personal data ethically and legally in Industry 5.0, where the collection, processing, and analysis of personal data are integral to smart environments. It focuses the importance of privacy, the ethical use of the data, and the necessity of adherence to such legal regulations as GDPR. One

Fig. 7.3 Protecting personal data in smart environments with GDPR guidelines

is presented as using a computer monitor where personal data are emphasized and a lock icon symbolizing data security and a scale iconizing the relationship between information privacy and the law.

7.5.3 Benefits of Ethical and Legal Data Handling

There are several benefits in adopting ethical and legal provisions on handling personal data of both the business establishments and the customers.

- **Compliance**: It is therefore possible to avoid excessive legal penalties by ensuring that organizations do not run the risk of non-compliance by adhering to regulations such as the GDPR and CCPA. The regulatory bodies are currently paying more attention to the companies in terms of infringing data protection and failure to comply with the data privacy provisions can subject companies to massive fines and legal wrangles. Observing privacy rules not only helps the businesses to avoid fines but also demonstrates the fact that the business is concerned with consumers privacy.
- **Trust**: Open practices in the management of data win the trust of the consumers. People tend to use technologies that collect their personal information when they are sure that their personal information is used ethically, is collected with their permission, and is stored safely. The industry of healthcare is especially sensitive, and trust is a key factor in such a case, as the privacy of data is a serious issue. Wearable devices, sharing personal health data, and engagement in digital health services are more likely to be used by consumers who believe that their privacy is respected.

- **Reputation**: Firms that are sensitive to ethical data operations enhance the corporate image of such a firm in the market. As the consumer is increasingly becoming conscious on the issue of their privacy and safety, they are increasingly being attracted to the company which has a very strong data protection policy in place. Maybe the most important aspect of data ethics is that adhering to them may be the difference between a company and its competitors and attract the customer who cares about his or her privacy [9].

7.6 Best Practices and Technological Safeguards of Mitigation Strategies

As Industry 5.0 is emerging, with data being produced faster than ever before and systems increasingly becoming more interconnected, firms are more vulnerable than ever to having their data breached, being the victim of a cyber attack, and privacy concerns. These risks can be least mitigated by ensuring that organizations adopt holistic mitigation strategy that will help integrate the best practices in cybersecurity, privacy policies and state of the art technological protection. These strategies are not only aimed at securing sensitive information but also building trust with the customers, fulfilling regulatory requirements and resiliency with business operations.

7.6.1 Best Practices of Cybersecurity

Cybersecurity cannot be left out in a risk management strategy of any organization. Introduction of IoT devices, AI systems, and connected machines has increased the attack surface and, therefore, exposed businesses to cyber threats. By implementing the best practices in cybersecurity, one can minimize the risk of an attack by securing the system, networks and data.

- **Zero Trust Architecture**: Zero Trust Architecture (ZTA) is a security model which considers the position that no device, user or system can be trusted, either within or outside the corporate network. Such an approach deploys tight measures of access and validation must always occur prior to any access to the network resources.

  Continuous Verification: Under the zero-trust system, there is an assumption that all users or devices are malicious until they can be demonstrated not to be. This encompasses verifying every user and device before they gain access to the resources, they are continuously tracked, and their identity is strictly checked such as the use of multi-factor authentication (MFA).

  Minimized Attack Surface: Zero Trust assists in minimizing the lateral movement inside the network of an organization. When a component of the system is breached, the attacker cannot have free range in the other sections of the network. That considerably minimizes the threat of mass violations and promotes the security of the entire network.

Example: In a smart factory, Zero Trust can be used to make sure that only authorized people and machines have access to sensitive systems of production. Even in the factory, in case a machine is hacked, or the credentials of a worker are stolen, Zero Trust will make sure that the person will be able to access only those areas of the system that they have permission to use.

- **Encryption**: One of the best methods of securing information is through encryption, and this information cannot be viewed by anyone, even when intercepted by the wrong people. Data being used on the internet or stored in databases needs to be encrypted so as to protect sensitive information.

  In Transit Encryption: The data transmitted are encrypted in transit; hence, the data is secure on the move through the networks. It also makes sure that data cannot be read even in case of interception unless the decryption key is used. Secure communications on the internet are secured using common encryption protocols like SSL/TLS.

  Data Encryption at Rest: Data encryption at rest is a technique that secures stored data, which means that information will be kept safe when it is not in the process of being sent. This is more so for sensitive data that is stored in the cloud or internal servers.

  Example: Within the healthcare industry, the data in an electronic health records (EHR) system should be encrypted in transit (e.g. during a transfer of data between healthcare providers) and at rest (e.g. in the database). This keeps away sensitive health information from being accessed by the unauthorized.
- **Frequent Security Audits**: Security auditing is among the preventative measures, which could be performed to identify the vulnerabilities of the systems, networks, and applications. These audits help establish the weak points in organizations which, subsequently, can be used by cybercriminals.
- **Vulnerability Scanning**: Vulnerability scanning, penetration testing and risk assessment are the common characteristics of security audits. With the help of scans, it is possible to identify the vulnerability of the settings in the system, outdated software, and other potential areas of vulnerability that can be leveraged against an organization.

  Patch Management: Security audits also help organizations to be aware of systems not patched or which have old software that also needs security patches. Through frequent updates, organizations are able to seal known loopholes as well as defend against new threats [10].

  Example: A smart city system that is integrated with IoT sensors and traffic management systems should be regularly audited in terms of security so as to detect any vulnerability that an attacker may take advantage of. Such audits are capable of contributing to the maintenance of systems that are current and able to resist cyber threats.

7.6.2 Technological Measures to Privacy of Data

Since Industry 5.0 technologies collect and process personal and sensitive information, the issue of privacy is gaining more and more importance. Technological safeguards may be used to assist the organization in securing personal information and guarantee that privacy laws are implemented.

- **Data Anonymization**: Anonymization of data is one of the methods of preserving personal information by stripping it of identifiable data. The anonymized data cannot be linked to a certain person, which also helps to preserve privacy but enables the business to examine the trends and patterns.

  Privacy Guarantee: Anonymization enables companies to analyze and optimize personal data without the risk of revealing their identities. It allows the companies to get meaningful information without intruding on privacy.

  Example: In a healthcare environment, anonymizing the data of the patients may enable the researchers to study the health trends of the population and not be able to identify the health status of the individuals. As an example, scientists will be able to examine the efficiency of treatment in various populations without revealing the patients.
- **Strong Encryption and Authentication**: This will give security to the sensitive data through encryption. End-to-end encryption will ensure safety of the data because it is communicated through the networks. Multi-factor authentication (MFA) adds one more level of protection where it is necessary to verify the identity of the user, and multiple techniques are used before gaining access to systems or sensitive information.
- **AI-Based Threat Detection**: The AI can be utilized to identify potential cyber threats and dangerous activity in real-time. The threats to the network can be detected and removed early enough before they can inflict significant damage to the system with the assistance of the analysis of the patterns of network traffic and identifying the deviation of this traffic form the norm.
- **Compliance and Privacy Protection**: The organizations need to ensure that they are informed of the new privacy regulations and are in compliance with them. The steps that are crucial to protecting consumer privacy are the use of data anonymization techniques, provision of consent of the user and transparency in using the data.
- **Safe Data Storage**: Data storage should be secured in order to protect the personal and sensitive data of the individuals against access by the non-authorised individuals. Organizations should ensure that they store data in safe environments like encrypted cloud storage and also provide access controls so that only authorized people or systems have access to the data.

Multi-Factor Authentication (MFA): In order to increase the security of the stored data, the organizations must implement MFA to access sensitive information. This provides another level of security, as the user is not allowed to select only one way of authenticating his or her identity. Instead, it has to be a combination of two or more, like a password and biometric data.

Access Control Mechanisms: The role-based access control (RBAC) mechanism will allow only individuals or systems to access the data that they require in their particular role, and hence, the risk of unauthorized access will be minimized.

Example: In a bank, the data of the customer accounts is located in an encrypted cloud environment with high access restrictions. The workers in customer service may access only the limited information related to their work, whereas access to the higher levels is restricted.

- **Blockchain for Privacy**: Blockchain technology offers a transparent, immutable, and secure history of accessing and making transactions with data. It can be employed to monitor the utilization of personal data to make sure that it is utilized or discussed for approved purposes only.

  Transparent Data Access: In the case of blockchain, all the activities concerning personal data are written on a decentralized ledger. This also makes it transparent since the access and use of data can be tracked to its origin by all the stakeholders. It also avoids manipulation of the data since nobody can tamper with the blockchain without reaching an agreement with the whole network [11].

  Example: Blockchain may be applied in supply chain management to monitor and validate the gathering and utilization of personal information, e.g. the preferences or the purchase history of a customer. This creates openness for consumers regarding the use of their data and guarantees that it is not used by unwarranted parties.

The two major aspects of digital security, visual comparison of two key aspects in digital security is presented in the Fig. 7.4: Best Practices for Cybersecurity and Technological Safeguards for Data Privacy. The best practices to ensure a secure use of the internet, the Zero Trust Architecture, Encryption, and Regular security Audits are set on the left side, where it is necessary to ensure that everything was verified continuously, the data was protected, and the vulnerabilities were observed. The Data privacy on the right side has technological Data privacy safeguards which include Data Anonymization, Data storage, and Blockchain privacy, which aim at safeguarding personal data, securing storage systems, and making data usage transparent and traceable.

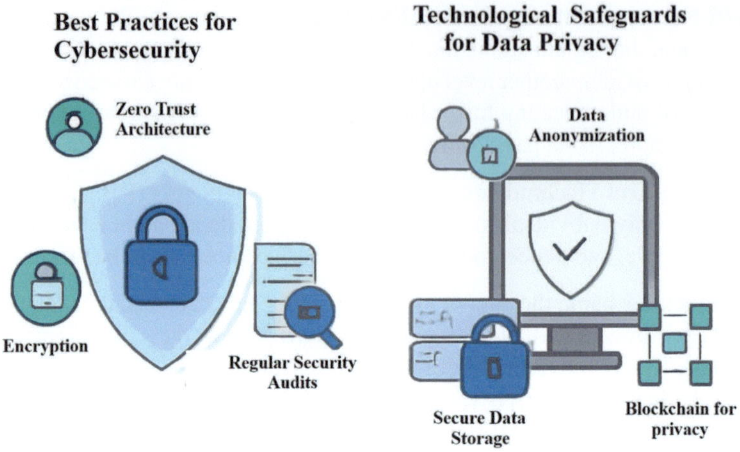

Fig. 7.4 Key strategies for cybersecurity and data privacy

7.6.3 Benefits of Mitigation Strategies

The benefits of effective mitigation strategies that will address both cybersecurity and data protection are numerous both to the organizations and consumers.

- **Data Security**: Zero Trust Architecture, encryption, and periodic security reviews are some of the best practices that will significantly reduce the risk of data breaches to ensure valuable information is not shared. These are just few measures towards securing the cyber and physical infrastructure.
- **Legal Compliance**: Mitigation practices can also help organizations to comply with data protection regulations, which become increasingly tougher each time, such as GDPR and CCPA. By observing these laws, the organizations are able to avoid the legal issues and penalties since they are certain that their organizations are upholding their responsibilities to ensure the privacy of the consumers.
- **Risk Management**: The proactive approach to cybersecurity and privacy will enable companies to anticipate and respond to the threats on time. This minimizes the threat of losses due to cyberattacks and reduces the interference with operations and loss of funds. Avoiding risks prior to their escalation will ensure that firms can remain in business and will not be hit by the adverse publicity that accompanies the occurrence of data breaches [12].

7.7 Analysis Perspective: Risk Assessment Frameworks and Breach Costs

As more advanced technologies are implemented in an organization in the guise of Industry 5.0, organizations will be prone to a number of cybersecurity and privacy threats. These risks can lead to severe financial, operation, and reputational losses when they are not managed appropriately. To overcome these threats, companies ought to come up with systematic risk assessment models that can enable them to identify their weak points and act accordingly against the threats. Moreover, an in-depth cost benefit analysis also allows organizations to approximate the monetary price of breaches, and it is because of this that they prioritize their investments in data protection and cyber security. This section is dedicated to the importance of risk assessment models, the cost of the violation, and the manner in which strategic risk management may reduce weaknesses and enhance resilience.

7.7.1 Risk Assessment Structures

Cybersecurity Risk assessment models are highly significant to organizations that desire to understand, examine and even mitigate against cybersecurity threats. These frameworks introduce a methodical manner of identifying vulnerabilities, identifying threats, reacting to an incident, and recovering the effect of a breach. Through the assistance of the well established frameworks, organizations are able to align their risk management processes to the industry standards, improve their security positions, and reduce their exposure to cyber risks.

- **NIST Cybersecurity Framework**: NIST Cybersecurity Framework is generally regarded as one of the most multifaceted and comprehensive risk management cybersecurity frameworks. Developed by NIST in collaboration with the industry professionals, the framework offers a set of recommendations that can be used by the organizations to manage and mitigate cybersecurity threats. The NIST framework is a universal industry, and any organization can use an adapted method of cybersecurity regardless of its large size.

  Identify: It is the first process that involves identifying significant assets like systems, data, and networks that have to be secured. The organizations also identify the weaknesses and threats that these assets might face either internally by other forces or by the weaknesses of the system.

  Protect: It is concerned with installing safeguards with the aim of protecting the identified assets. This includes access control, employee training, encryption, and firewalls among other technical controls that minimize the risks before they take place.

Detect: This is significant to identify cybersecurity incidents with the view of minimizing the impact of potential breaches. This is accomplished through the development of the real time monitoring and anomaly detection systems which help in the process of detecting suspicious activities in the network, systems, and devices.

Respond: Organizations should have a response plan in case of breach or incident in the cybersecurity. These involve identification of the origin of the attack, capturing the intrusion and averting the loss on the spot.

Recover: The final stage is the recovery in which the organizations restore systems and data to the normal working state. The stage also involves post-incident analysis in order to understand what was wrong and how to improve cybersecurity in future.

The NIST cybersecurity framework allows the organizations the liberty to adapt to the evolving cybersecurity concerns. Following its principles, the businesses will be capable of managing risks and be better prepared to act upon security incidents.

- **ISO/IEC 27001**: Standard on information security management ISO/IEC 27001: ISO/IEC 27001 is an accepted standard all over the world. It outlines the ideal practices in the creation, setting up and upkeep of an information security management system (ISMS) to protect sensitive information. This model helps an organization in quantifying risks and initiating control measures that can maintain security measures to be improved at any given time.

  Risk Assessment: ISO/IEC 27001 also emphasizes on the need to have a methodical procedure of risk assessment, so as to determine the potential threat to information security. The standard also requires the organizations to assess the likelihood and aftermath of such risks so that to determine the controls and safeguards.

  Controls and Safeguards: The framework provides a guideline of implementing some security controls to prevent the identified risks, which are access management, physical security, encryption, and incident management.

  Continuous Improvement: ISO/IEC 27001, encourages continuous improvement of the ISMS of the organizations as in form of frequent audits and revisions. This is a proactive approach in ensuring that security is in check and ahead of the new and emerging threats.

  With the ISO/IEC 27001, organizations are able to show that they are committed to the protection of sensitive information, abide by the data protection laws, and earn customer and stakeholder trust [3].

7.7.2 Expenses

Though the implementation of risk assessment frameworks is linked to the elimination of data breaches and minimization of cybersecurity risks, it should be noted that organizations should understand the extent to which the costs of data breaches might be. Cybersecurity breaches can be quite expensive, and it can be both monetary and indirect (the loss of reputation and operational disturbance).

- **Financial Losses**: The consequences of a data breach may be huge, such as legal consequences, fines, and claims. These expenses may multiply rapidly, particularly in case the violation includes sensitive data of customers, including payment information or individual medical records.

  Penalties and Legal Fines: Businesses can be fined heavily in case of non-compliance with data protection laws such as General Data Protection Regulation (GDPR) or California Consumer Privacy Act (CCPA), depending on the degree of the breach and the jurisdiction. To illustrate, a company can be fined up to 20 million euros or 4% of the annual world turnover, whichever is greater, under GDPR.

  Compensation Claims: Firms can also be subject to pay compensation to people whose personal information was disclosed or abused. It may result in class-action lawsuits and other expenses. Where there is a breach of sensitive medical data, such as in the case of a breach of medical data, there is a possibility that healthcare providers will be obligated to pay the affected patients' compensation as a result of the exposure of their medical data.

  Example: In 2017, Equifax, a consumer credit reporting agency, was one of the victims of a huge data leak that involved around 147 million individuals. The Federal Trade Commission (FTC) charged the company a fine of 575 million dollars to come to terms with the claims made against it, and compensation to the affected people just increased the financial strain [13].
- **Reputation Damage**: In addition to the actual monetary losses, breaches may have a devastating effect on the reputation of a company. In cases where the business fails to secure the data of the consumers, trust is destroyed, and chances of the consumers interacting and purchasing their products again are reduced.

  Loss of Customer Trust: When there is a breach in terms of personal or financial information, it may cause a loss in customer trust, and this may not be easy to recover. With the ever-growing competitive market today, consumers are now more privacy-conscious, and they expect businesses to ensure that their details are secured. A track record of violations may push away customers from the competitors, resulting in a loss of market share.

  Brand Reputation: In addition to the trust of the customers, the violations can also have an adverse impact on the general brand reputation of a company. Negative media coverage and flashbacks on social media do not help the breach since the company would be left scratching its head to get back on its feet.

  Case Study: In 2013, the Target data breach traced and stole the personal and payment card information of 40 million customers. The hack did not only cost the company money but also hurt the brand of Target and the shoppers were less likely to spend money in the store. The company had to spend millions of dollars on the restoration process and had to lose in the long-term perspective in reputation.

- **Operational Disruption**: A security failure in the cyber world can cause severe operational effects particularly when critical systems, data or services have been compromised. Such interruptions may result in a long period of downtime, reduced productivity, and high recovery expenses.

  Downtime of the System: In case of a cyberattack on the key systems or data, companies might have to shut down operations until they seek the origin of the attack and limit its consequences. This dead time may cause late delivery, missed targets, and wastage of business opportunities.

  Recovery Costs: There are costs of recovery besides downtime, which organizations have to spend on recovering a breach. This may involve expenses incurred on repairing the system, restoring information and measures put in place to ensure that the system is not attacked again [14].

  Example: WannaCry ransomware of 2017 infected more than 200,000 systems in 150 countries, including such large organizations as the National Health Service (NHS) in the UK. The hack generated much havoc, and the healthcare providers could not access patient records, and even appointments were cancelled. The NHS projected the economic cost to be 92 million, covering the costs of recovery and disruption of operations.

Figure 7.5 demonstrates a professional in the situation of risk assessment in the area of cybersecurity, focusing on the significance of risk structures and possible expenses of intrusions. Among the main risks represented are financial losses, reputation damage, and operational downturn, which demonstrates the necessity of the active approach.

Fig. 7.5 Evaluating cybersecurity risks and costs

7.7.3 *Benefits of Risk Mitigation Strategies and Risk Assessment Framework*

There are numerous benefits related to the use of risk evaluation frameworks and controls because it allows the organization to make more effective decisions regarding their investments in cybersecurity and privacy and reduces the cost of breaches in monetary and reputational terms [15].

- **Informed Decision-Making**: Risk assessment frameworks may provide an organization with information that is based on data, which can help in prioritizing the investment in cybersecurity in terms of the risks and expenses involved. The businesses can also better direct the resources on the protection of the valuable assets with the knowledge of the vulnerabilities and the impacts of the breach.
- **Less Financial Losses**: Risk management through risk evaluation and mitigation prior to occurrence, assists a business to prevent breaches and decrease the monetary cost of cybersecurity incidents. Threats can be identified and acted upon early to prevent massive destruction and reduce the costs required to recompense the victims, pay legal fees, and restore the image.
- **Better Risk Management**: The vigorous risk management process enables the organization to be proactive of the emerging threats and become resilient in managing cyber risk. Through the processes of constant assessment of vulnerabilities, controls, and the improvement of incident response plans, companies can become more capable of responding to incidents quickly and possibly reduce the damage that can be done [16].

To sum up, industry 5.0 relies on data analytics, cybersecurity, and privacy. With the adoption of advanced analytics, businesses can be in a better position to make better business decisions, improve operations, and have a competitive edge. They should, however, be keen on securing their systems and sensitive information against cyber-attacks and privacy rules. With the further development of the digital environment, the necessity of a strong data management strategy and security measures will be even higher.

7.8 Learning Outcomes

After reading this chapter, the readers will be in a position to:

- Learn how industry 5.0 has led to the growth of data analytics to prescriptive and predictive analytics and how this will affect decision-making.
- Determine the major cybersecurity threats of IoT, industrial control systems, and human error and share the measures to mitigate the threats.
- Learn about the ethical and legal aspects of processing personal data in Industry 5.0 and consider the ways to make sure that the provisions of privacy legislation such as GDPR and CCPA can be met.

- Assess the performance of the best cybersecurity practices such as zero-trust architecture, encryption, and frequent security audit in securing digital systems and data.
- Use risk assessment models such as NIST and ISO/IEC 27001 to determine the threats to cybersecurity and the financial implication of data breach on organizations.
- Establish an end-of-end strategy in dealing with cybersecurity risks, protection of connected systems, and consumer confidence in the 5.0 industry setting.

7.9 Review Questions

1. (a) What is the change in descriptive to prescriptive analytics in Industry 5.0?

   (b) What is the role of prescriptive analytics in decision-making?
2. (a) Identify and study the risks of cybersecurity related to the use of Industry 5.0 of IoT devices and industrial control systems.

   (b) What should be done to ensure that such systems are not attacked by hackers and how can sensitive information be secured?
3. (a) Write about the privacy issues that will be related to the gathering and utilization of personal information in Industry 5.0 and especially the smart environments?

   (b) What are some of the ways through which businesses can make the process of data collection ethical and legal in accordance with laws such as GDPR and CCPA?
4. (a) Which are the main elements of a Zero Trust security model, and how can it reduce the risks of cybersecurity in Industry 5.0?

   (b) Give the examples of the Zero Trust architecture implementation in the smart manufacturing environment.
5. (a) How will data anonymization and encryption help in improving data privacy in Industry 5.0?

   (b) Briefly describe the technological measures that can be used by the companies to secure sensitive data, and how the blockchain can guarantee data transparency and integrity.
6. (a) Assess the place of predictive maintenance in Industry 5.0 and how it may affect the efficiency of operations.

   (b) What are the benefits of the predictive maintenance process with digital twins and AI-driven analytics and how do they work?
7. (a) Evaluate the economic cost of cybersecurity violation to organizations. How much can be lost with a data breach, what are the financial losses, reputational losses, and operational losses?

(b) What are the ways through which businesses can quantify and reduce such expenses?

8. (a) Explain the role of the models of risk assessment like NIST Cybersecurity Framework and ISO/IEC 27001 in the process of determining and managing cybersecurity threats.

   (b) What are some of the ways through which organizations can leverage these frameworks to better their cybersecurity position under Industry 5.0?
9. (a) What are the ways in which competitiveness of businesses is improved by the application of data-driven decision-making in Industry 5.0?

   (b) Explain the role of advanced analytics and real-time data processing in making better and efficient decisions in such industries as manufacturing and logistics.
10. (a) How can companies make sure that they adhere to the environmental and data protection laws when operating in Industry 5.0?

    (b) What is the role of the regulatory standards such as the GDPR and CCPA regarding the processing of personal data in the connected environment?

Appendix: Solution Manual

1. **(a) What is the change in descriptive to prescriptive analytics in Industry 5.0?**

Solution: Prescriptive analytics, on the other hand, uses advanced algorithms to recommend actions based on predictions.

(b) What is the role of prescriptive analytics in decision-making?

Solution: It allows businesses to not only forecast trends but also optimize processes, allocate resources efficiently, and make proactive decisions. The benefits include enhanced decision-making, improved operational efficiency, and a competitive edge in identifying trends before competitors.

2. **(a) Identify and study the risks of cybersecurity related to the use of Industry 5.0 of IoT devices and industrial control systems**

Solution: Cybersecurity risks include unauthorized access, ransomware attacks, and data breaches. IoT devices, being connected to various systems, are vulnerable due to inadequate security features.

(b) What should be done to ensure that such systems are not attacked by hackers and how can sensitive information be secured?

Solution: To secure these systems, businesses must implement strong access controls, ensure regular patching and updates, and use AI-based anomaly detection systems.

Encryption and secure industrial control systems (ICS) are essential for protecting sensitive data and ensuring system integrity.

3. **(a) Write about the privacy issues that will be related to the gathering and utilization of personal information in Industry 5.0 and especially the smart environments?**

Solution: Privacy concerns arise from the extensive collection of personal data, such as employee health metrics or customer behavior.

(b) What are some of the ways through which businesses can make the process of data collection ethical and legal in accordance with laws such as GDPR and CCPA?

Solution: Businesses must ensure that data collection is based on consent, anonymize personal data, and comply with privacy regulations like GDPR and CCPA. Legal requirements include the protection of user rights, transparency in data usage, and ensuring that data is not misused or sold without consent.

4. **(a) Which are the main elements of a Zero Trust security model, and how can it reduce the risks of cybersecurity in Industry 5.0?**

Solution: Zero Trust architecture assumes that no entity, whether inside or outside the network, can be trusted by default. It requires continuous authentication and verification before granting access.

(b) Give the examples of the Zero Trust architecture implementation in the smart manufacturing environment.

Solution: In smart manufacturing, this could mean that each device and worker is constantly monitored, with access granted only after authentication. This minimizes the risk of internal and external threats.

5. **(a) How will data anonymization and encryption help in improving data privacy in Industry 5.0?**

Solution: Data anonymization ensures that personal identities are not exposed while still allowing data analysis. Encryption protects data in transit and at rest, ensuring it is unreadable to unauthorized users.

(b) Briefly describe the technological measures that can be used by the companies to secure sensitive data, and how the blockchain can guarantee data transparency and integrity.

Solution: Blockchain can be used to track data access and modifications, ensuring transparency and securing sensitive information from tampering.

6. **(a) Assess the place of predictive maintenance in Industry 5.0 and how it may affect the efficiency of operations.**

Solution: Predictive maintenance uses data analytics and machine learning to predict when equipment will fail, allowing for timely interventions.

(b) What are the benefits of the predictive maintenance process with digital twins and AI-driven analytics and how do they work?

Solution: Digital twins provide real-time models of physical assets, enabling businesses to simulate and predict equipment performance. This approach reduces downtime, extends equipment life, and optimizes resource allocation.

7. **(a) Evaluate the economic cost of cybersecurity violation to organizations. How much can be lost with a data breach, what are the financial losses, reputational losses, and operational losses?**

Solution: The financial impact of a cybersecurity breach includes direct costs (e.g., fines, legal fees) and indirect costs (e.g., loss of customer trust, reputational damage). Operational disruption, such as downtime, can also result in significant losses.

(b) What are the ways through which businesses can quantify and reduce such expenses?

Solution: Businesses can mitigate these costs by implementing proactive security measures, conducting regular security audits, and having a robust incident response plan in place.

8. **(a) Explain the role of the models of risk assessment like NIST cybersecurity framework and ISO/IEC 27001 in the process of determining and managing cybersecurity threats.**

Solution: The NIST Cybersecurity Framework helps organizations identify vulnerabilities, protect against threats, detect anomalies, respond to incidents, and recover from breaches. ISO/IEC 27001 provides a systematic approach to managing sensitive company information.

(b) What are some of the ways through which organizations can leverage these frameworks to better their cybersecurity position under Industry 5.0?

Solution: Both frameworks guide organizations in building comprehensive cybersecurity systems that enhance their resilience to cyber threats.

9. **(a) What are the ways in which competitiveness of businesses is improved by the application of data-driven decision-making in Industry 5.0?**

Solution: Data-driven decision-making enhances competitiveness by allowing businesses to make **timely, informed decisions based on real-time data.**

(b) Explain the role of advanced analytics and real-time data processing in making better and efficient decisions in such industries as manufacturing and logistics.

Solution: In manufacturing, for example, real-time analytics can optimize production schedules, reduce waste, and improve supply chain efficiency, leading to cost savings and better customer satisfaction.

10. **(a) How can companies make sure that they adhere to the environmental and data protection laws when operating in Industry 5.0?**

Solution: To ensure compliance with regulations like GDPR and CCPA, businesses must implement robust data protection measures such as encryption, access control, and data **anonymization**.

(b) What is the role of the regulatory standards such as the GDPR and CCPA regarding the processing of personal data in the connected environment?

Solution: Regular audits, clear data handling policies, and transparency with consumers regarding their data rights are essential for maintaining compliance in connected environments.

Bibliography

1. Domenteanu A, Cibu B, Delcea C (2024) Mapping the research landscape of Industry 5.0 from a machine learning and big data analytics perspective: a bibliometric approach. Sustainability 16(7):2764
2. Ingard A (2025) The critical role of data utilization in achieving sustainable success for medium-sized manufacturing SMEs in Pathum Thani Province, Thailand. Sustain Fut Life Cycle Asses Ind 2025:403–422
3. Hassan MA et al (2024) Systematic analysis of risks in Industry 5.0 architecture. Appl Sci 14(4):1466
4. Hammad M et al (2025) Countermeasures to tackle the security challenges in smart manufacturing. In: Smart manufacturing blueprint: navigating industry 4.0 across diverse sectors. Springer, Cham, pp 375–409
5. Khan N et al (2024) Explainable AI-based intrusion detection system for industry 5.0: an overview of the literature, associated challenges, the existing solutions, and potential research directions. arXiv preprint arXiv:2408.03335
6. Santos B, Costa RLC, Santos L (2024) Cybersecurity in industry 5.0: open challenges and future directions. In: Proceedings of the 2024 21st annual international conference on privacy, security and trust (PST). IEEE
7. D'Acquisto G et al (2015) Privacy by design in big data: an overview of privacy enhancing technologies in the era of big data analytics. arXiv preprint arXiv:1512.06000
8. Rehman A et al (2022) A secure healthcare 5.0 system based on blockchain technology entangled with federated learning technique. Comput Biol Med 150:106019
9. Hammad M et al (2025) Heavy industry and machinery: building resilience with smart manufacturing. In: Smart manufacturing blueprint: navigating industry 4.0 across diverse sectors. Springer, Cham, pp 289–321
10. Siew E-G et al (2025) Tackling the human element in cybersecurity management: a tale of a phishing attack on a Malaysian fashion company. J Inform Technol Teach Cases 2025:20438869251360810
11. Yuhala P et al (2025) On real-time guarantees in intel SGX and TDX. In: Proceedings of the 37th Euromicro conference on real-time systems (ECRTS 2025). Schloss Dagstuhl–Leibniz-Zentrum für Informatik
12. Karaja MB et al (2024) AI-driven cybersecurity: transforming the prevention of cyberattacks
13. Wu S (2021) Asian technical experts: skilled corporate migration in the global commodity chain. Global Netw 21(2):413–433
14. Hammad M et al (2023) Framework for the implementation of smart manufacturing systems: a case in point. Processes 11(5):1436

15. Kasula VK et al (2024) Fortifying cloud environments against data breaches: a novel AI-driven security framework. World J Adv Res Rev 24:1613–1626
16. Aiello S (2025) Prescriptive zero trust-assessing the impact of zero trust on cyber attack prevention. arXiv preprint arXiv:2508.12953

Chapter 8
Industry 5.0 in Different Sectors

8.1 Objectives

- To analyze how industry 5.0 is going to transform the manufacturing, healthcare, agriculture, and smart cities, focusing on the combination of modern technologies with human interaction.
- To understand how the industry 5.0 technologies can allow mass customization and collaborative production line in a manufacturing industry.
- To examine the role of industry 5.0 in improving the personalized medicine, robotic surgery, and patient monitoring in healthcare.
- To address the issue of precision farming, sustainable resource use and integration of IoT in agriculture.
- To determine the role of smart infrastructure, energy management and citizen engagement in the creation of smart cities.
- To determine the barriers and enablers that affect the implementation of the industry 5.0 technologies in different industries and regions.

8.2 Scope

The fifth generation of industry (Industry 5.0) will be radically transformative of a lot of industries, and it will be a mixture of the newest technologies, e.g., artificial intelligence (AI), Internet of Things (IoT), robotics, and cyber-physical systems, and the creativity, skills, and cooperation of humans. The new paradigm is the break of the previous paradigm of automation-based methods, and it emphasizes the role of human involvement, collaboration, and innovation in driving productivity and innovation across the industries. The future of industry 5.0 will be more human based on production, service delivery, and solving problems to make it more personal, sustainable, and efficient in different areas. As the industries are transforming due to a

M. Hammad et al., *Industry 5.0: The Human-Centric Future of Industry*, Springer Series in Advanced Manufacturing, https://doi.org/10.1007/978-3-032-12622-1_8

more connected and smarter ecosystem, the part of the human worker will also change along with the machines, enhancing creativity, flexibility, and decision-making.

The revolutionary aspect of Industry 5.0 in four key domains, that is, manufacturing, healthcare, agriculture, and smart cities, are discussed in the section below. In both of the sectors, we will examine the applications of each field which are imperative to the industry, the problems that they incur, and the efficiency, sustainability, and the better human–machine interaction advantages that they provide.

8.3 Manufacturing: Customization and Co-operative Production Lines

Historically, manufacturing is one of the main fields of interest in the industrial revolutions. The introduction of the industry 5.0 has transformed the manufacturing industry into a spectacular one that not only incorporates the innovative technologies, including robotics, AI, and digital twins but also the further consideration of the human creativity and the dynamics of teamwork. This change signals the death of the old-fashioned method of production which mostly focused on mass production, mass production, and standardization of production process to more flexible, human and personalized production processes.

8.3.1 Customization

One of the key ideas of the revolution of the industry 5.0 is mass customization. Unlike mass production where the products are manufactured in large volumes with little or no chances of customization, Industry 5.0 allows the manufacture of highly customized products in large volumes. The powerful combination of AI, robotics, and data analytics allows manufacturers to create products of particular shape, color, size, etc., that would suit the particular preferences and needs of individual consumers, and it would not reduce the efficiency and cost-effectiveness of the production process because its scale. Under this paradigm shift, there is a high degree of customization, and this customization could be in the form of bespoke consumer electronics and customized fashion so that not only are the products of high quality, but also in tune with the special requirement of individual consumers.

Advanced Robotics to Personalization: Collaborative robots or cobots are the key to the mass customization in manufacturing. The robots are applicable in conjunction with the human operators who are not substituted by them. They are used in their place of complementing the capabilities of human operators. To illustrate the point, in the industries where the products are manufactured, e.g., automotive manufacturing, cobots assist in assembling their parts, inspect them, and offer a certain quality control. In the same way, in the industry like fashion, cobots can cooperate with human designers to produce clothes that are customized to the size of specific customers using such technologies as 3D knitting or embroidery machines. This has put companies in a position to respond to the market fast by their production lines and manufacture more specialized products according to the expectations of individual consumers.

AI-Powered Personalization: The data on consumers is analyzed through machine learning to predict and develop personalized designs of products. AI systems allow the consumer electronics market to design products that can fit the specific preferences of the users, adjusting such parameters as the size of the screen, the performance features, and the design elements according to the information they get. This allows the companies to sell their products which are more suitable to the needs of narrow markets or to a particular set of customers, which supplements the consumer's experience [1].

8.3.2 Collaborative Production Lines

One of the most significant features of Industry 5.0 in manufacturing is the idea of collaborative production lines, where humans and robots work in the same environment without any problems. In conventional manufacturing, machines were usually separated from the human worker, where each of the machines would be engaged in different activities along the production line. Conversely, the collaborative production lines enable a more fluid combination of human skills and machine capabilities, forming a symbiotic relationship which increases overall productivity, flexibility, and quality.

Human–Machine Collaboration: In the collaborative production lines, human laborers and robots occupy the same working environment and collaborate on the same tasks, including assembly, packaging, and quality control. The cobots help in doing boring and strenuous work, e.g. lifting heavy parts or assembling complex parts, whereas humans do higher-level jobs, which involve problem-solving, critical thinking and design input. The possibility to distribute the workspace between machines and people creates a more productive and peaceful atmosphere in the production process, and every agent makes a contribution to the success of the resultant product.

AI and Machine Learning: AI-based systems are being added to the production lines to watch over the processes, foresee breakdowns of the equipment, and streamline production plans. The AI tools may be trained in the behavior of people, and they enhance with time; thus, production lines are always in the process of improvement to make them more efficient. Digital twins: Virtual models of physical production lines Digital twins are also an extension of this integration because they allow monitoring and simulation of the production process in real time. It allows the manufacturers

Fig. 8.1 Human–robot collaboration drives customized production in Industry 5.0

to see possible bottlenecks, optimize the resource allocation and make sure that the whole system operates at optimum [2].

Figure 8.1 depicts the modern manufacturing setting where Industry 5.0 is showcased through the collaboration between human workers and robots. The female worker is engaged in customizing a product, with the assistance of a robotic arm, in a smart factory designed for mass customization. Advanced technology, including a digital interface, enhances human capabilities and allows for tailored production processes.

8.3.3 Benefits

Numerous advantages accompany the use of Industry 5.0 in manufacturing, both for the business, the workers, and the consumers. These are greater flexibility, better quality control and more safety for the workers.

- **Greater Flexibility**: One of the most important benefits of Industry 5.0 is the possibility to easily adjust production lines to shifts in customer demand or changes in product designs. With the help of AI, robotics, and IoT technologies, manufacturers can now operate flexible production lines and create batch sizes as small as they want or specialize their products entirely without losing efficiency. This enables companies to satisfy the various needs of consumers and also achieve a competitive edge in the market.
- **Better Quality Control**: AI and real-time data analytics integration in the manufacturing process enhances quality control because the defects are identified early

in the production process. AI systems are able to track the performance of single machines, predict possible failure, and take corrective measures prior to them leading to expensive mistakes. Digital twins offer an opportunity to perform testing and optimization of the production process in a virtual environment, guaranteeing the production of the final product of the highest quality and reliability.

- **Increased Safety of Workers**: The teamwork of Industry 5.0 is a better way to increase the safety of the workers because it transfers risky, dull or physically intense jobs to robots and cobots. This decreases the chances of accidents in the workplace, like repetitive strain injuries or even working in harmful environments, so that human workers can concentrate on more intellectual tasks. Consequently, employees feel more satisfied, involved, and healthy at work, which leads to a better culture in the workplace [3].

Overall, the application of the technologies of Industry 5.0 in the manufacturing sector is a paradigm shift in designing, developing, and providing products. Industry 5.0 is giving manufacturers an opportunity to offer high-quality products of a truly personalized type and be efficient, flexible, and safe due to the integration of human creative and technologically innovative forces [4].

8.4 Healthcare: Personalized Medicine, Robotic Surgery, and Patient Monitoring

The healthcare sector is facing a radical change due to Industry 5.0 and its human attitude. An amalgamation of new technologies, such as artificial intelligence (AI), robotics, the Internet of Things (IoT), and human talents is in its path to creating very personalized, very efficient, and very effective care. The revolution is not only enhancing the outcome of patients but also streamlining the operations and making the risks involved in the traditional form of healthcare practices much lower. This will, therefore, allow caregivers to become more precise, timely and tailored in their provision of healthcare services and in the process improve a superior union between the patient and the healthcare practitioners. Personalized medicine, robotic surgery and high-level patient monitoring are some of the biggest Industry 5.0 interventions in healthcare and it is through these interventions that the expectations of patient care are redefining.

8.4.1 Personalized Medicine

Personalized medicine is one of the industries in healthcare where Industry 5.0 is making significant advances in this field. The shift of the mass model of therapy towards the individualization of the treatment choice based on the specifics of each individual, such as genetic, lifestyle, environmental factors, etc., is increasingly enabled with the assistance of AI and data analytics. In Industry 5.0, artificial intelligence algorithms process a large amount of patient information to generate an exclusive treatment plan that would ideally fit the requirements of a specific patient and include their genetic history, family medical history, lifestyle, and environmental conditions. This will facilitate administration of more effective treatment with few side effects that will improve the quality of care.

AI-Guided Individual Treatment Plans: With the help of AI in dealing with big and complicated data the healthcare sector is changing the manner of treating patients and how they are diagnosed. The treatment of diseases like cancer, diabetes and heart disease in the past was normally a generalized approach that may not take place in all patients. However, today, when AI is implemented, medical staff members can also design the treatment, which can be tailored to the genetic predisposition of the patient and other personal characteristics. Taking the example of AI, it can interpret the genetic data of a tumor of a patient and determine the mutations, and recommend the most effective type of targeted treatment that can potentially raise survival rates and reduce the number of side effects.

Case in Point: AI in Oncology: Personalized medicine is also advancing in the oncology field. The AI-based algorithms are capable of analyzing a profile of the tumor of a specific patient, genetic indicators, and mutations to recommend a targeted therapy or a combination of drugs that ought to be used. A case in point is the case of genomic information that has been used to identify some of the mutations in breast cancer patients that have led to more effective treatment including that of the HER2-targeted therapy where the HER2 protein that makes cancerous cells grow is specially targeted and inhibited. Such accuracy minimizes the possibility of side effects and enhances the treatment results.

Pharmacogenomics: Pharmacogenomics is a field of personalized medicine that deals with the way the genetic makeup of an individual affects the response to any medication. Pharmacogenomic data are being analyzed with the help of AI to determine the most efficient medication depending on the genetic profile of an individual. This will help doctors choose the best medications to give to the patients, hence reducing the trial-and-error method adopted in traditional medication [5].

8.4.2 Robotic Surgery

The other field in which Industry 5.0 is having a dramatic effect on healthcare is robotic surgery. Robotic systems with AI are increasing the accuracy, control, and minimally invasive nature of surgeons, and this means that even complicated procedures are being carried out more accurately and with less risk. An example Case in point: AI in Oncology.

With the help of such robotic systems, the surgeon can do his or her operations by making smaller incisions, which result in faster recovery, reduced blood loss and postoperative pain by the patient. Besides that, more and more robotic surgery systems are integrated with real-time data and digital twins, making them more efficient.

AI-Based Accuracy and Low-Impact Procedures: Robotic surgery devices are designed to assist surgeons by providing better dexterity and high-definition and accuracy in 3D images that are not possible by human hands. An example is one of the most successful robotic systems the Da Vinci Surgical System, which provides a surgeon with a chance to perform delicate surgery, as prostatectomy, with enhanced control. This system uses robot arms that are controlled by a surgeon at a console during the operation and offers greater precision and fewer tremors which is critical when it comes to complications minimization in the surgery procedure.

Case in Point: Robotic-Assisted Cardiac Surgery: Robotic systems are also applied in cardiac surgeries to conduct minimally invasive surgeries on the heart. The systems enable the surgeons to operate bypass surgeries or valve replacement using very small incisions that minimize the chances of infection, less recovery period and less trauma to the body of the patient. With the help of robotic surgery, real-time imaging, and AIs to guide the surgeon, the precision with which the surgeon can operate is nowadays far greater than in the era of open-heart surgery.

Digital Twins and Real-Time Data: Another thing which Industry 5.0 is associated with is digital twins, which are virtual models of a patient's anatomy. Surgeons may plan and rehearse complicated procedures in the virtual environment using such detailed scans as CT scans or MRIs, and then perform them on the patient. This helps in reduction of errors, optimization of the surgical outcomes and reduction of the risks of complications that may occur in the course of the surgical operation. Counting on AI implementation in such systems, the surgeon has real-time navigation in the surgical process and is assisted in making accurate movements due to specifics of the patient [6].

8.4.3 Patient Monitoring

Patient monitoring is another crucial sector that Industry 5.0 is breaking into in the healthcare sector. With the use of AI-enabled analytics, IoT-enabled devices and wearable sensors, it is becoming possible to monitor patient vitals in real-time and 24 h a day, both at a hospital and at home. These gadgets provide direct access to vital medical information by the medical practitioners and can provide proactive care and prevent occurrence of health issues before they become serious diseases. This would be of particular help to the chronically ill patients, having chronic conditions such as diabetes, heart diseases, respiratory illnesses among others and who require constant attention and modifications in their treatment regimen.

IoT-Enabled Wearables in the Management of Chronic Diseases: IoT Wearables in chronic disease management: The application of wearable devices that capture vital signs such as heart rate, blood pressure, oxygen, and glucose levels in the management of chronic diseases is getting to be the rule. The devices record continuous measurements that are transmitted to the healthcare providers in real-time to assist them in checking the condition of the patient without necessarily visiting the patient frequently. As an example, diabetic patients can use continuous glucose monitors (CGMs) that indicate the amount of sugar in the blood throughout the day and warn the patient and their doctor in case the sugar level was either excessively high or excessively low. This will help them make the necessary changes to medications in good time to improve the disease management.

Remote Monitoring and Telemedicine: IoT-based remote monitoring features are also driving the boom in telemedicine whereby patients can receive medical consultation and advice remotely. To take one example, AI-based systems can analyze the data gathered by wearables on patients and give certain information to be considered by the physicians remotely. This particularly applies to elderly patients or patients who live in remote locations where medical services may not be readily available to them. Moreover, the AI algorithms are able to identify patterns and anomalies of the data and alert medical workers about the potential issues that may require immediate resolution.

Continuous Vital Sign Monitoring: IoT is also being implemented in hospitals where patients who need constant attention are checked with the help of vital signs. These devices can be used to measure numerous parameters such as temperature, pulse rate, respiratory rate, and oxygen saturation. The AI systems process the data ensured by these devices and could detect tendencies and anticipate any negative outcomes, including cardiac arrest or respiratory breakdown, so that the medical personnel would be able to intervene in time to rescue the patient [7].

The Fig. 8.2 demonstrates the application of AI and robotics in healthcare and in an Industry 5.0 environment, it presents personalized medicine, robotic surgery, and patient monitoring. The medical practitioner, wearing scrubs, interprets patient information with the help of a tablet, with the robotic arm doing surgery, and the focus on how human knowledge and new technologies, such as robotics and AI, can work together to deliver personalized care and enhance patient outcomes. There is also a digital display with the medical data of the patient displayed on the background.

8.4.4 Benefits

The application of Industry 5.0 in healthcare has many advantages that improve patient outcomes and operational efficiencies.

- **Better Patient Outcomes**: There has been an increase in more effective and individual patient-specific treatments because of the shift to personalized medicine and robotic surgery. The individualized treatments lead to limited side effects and increased rates of recovery. Robotic surgery is very precise and less invasive,

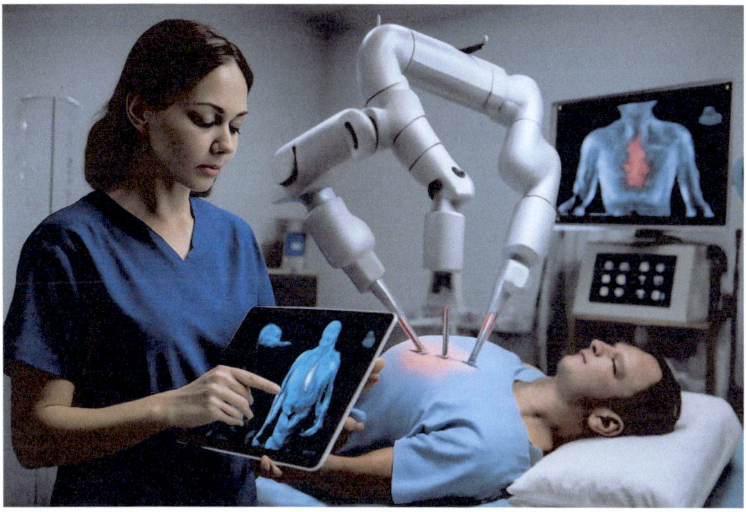

Fig. 8.2 AI-powered robotic surgery enhances precision and patient outcomes

which greatly decreases the risk and ensures faster recovery, which leads to higher overall patient satisfaction.

- **Lower Expenses**: Automation and AI in the diagnosis, monitoring of the patient, and planning treatment can lower the costs of healthcare by simplifying repetitive processes and decreasing the number of mistakes. Administrative tasks, including patient scheduling and billing, can be automated with the help of AI-powered systems so that healthcare professionals can concentrate on high-impact tasks. Moreover, the constant observation provided by IoT devices will allow avoiding expensive complications as the problems will be identified early, and treatment may not be so expensive in the future.
- **Empowered Patients**: As wearable sensors and remote monitoring systems are in development, patients have become more empowered in the process of managing their health. With these technologies, the patients are able to monitor their vitals and condition on a real-time basis and thus have improved self-management of chronic diseases. Moreover, patients will be able to get feedback from healthcare providers instantly, which will increase their feelings of control over their health and create a more collaborative relationship between the patient and his/her medical team [8].

Finally, it is possible to note that Industry 5.0 is revolutionizing the healthcare industry by making it more personal, efficient, and effective. The combination of AI, robotics, IoT, and human knowledge is leading to a future where healthcare will be more personal, recovery will be quicker, and operation costs will be lower. The more the sector adopts these advancements, the more patients and medical practitioners will enjoy responsive, data-based and more holistic care solutions.

8.5 Agriculture: Precision Farming and Sustainable Use of Resources

Agriculture has always remained an important sector which provides food security and sustains the economies of the world. However, with the world population increasing, the conventional ways of farming are under pressure to produce more food to satisfy the growing population and, at the same time, solve environmental problems. In its turn, Industry 5.0 is transforming agriculture with the help of the latest technologies such as IoT sensors, artificial intelligence (AI), drones, and robotics that allow farmers to maximize the production of crops, save resources, minimize waste, and increase food security. This move towards more efficient, sustainable and data-driven farming is referred to as precision farming and sustainable resource use.

8.5.1 Precision Farming

Precision farming is one of the aspects of Industry 5.0 in agriculture, which entails the application of state-of-the-art technologies to gather and analyze real-time data on different sources in an attempt to streamline farming activities. Through data-based information, farmers will be in a better position to determine the best practices to adopt in crop management, optimal inputs and enhanced productivity. This practice increases crop productivity not only but also minimizes the waste of resources in farming and makes it more efficient and sustainable.

- **IoT Sensors to Collect Data in Real Time**: The field IoT sensors may be applied to monitor the vital variables (soil conditions, temperature, humidity, weather conditions and crop health) constantly. These sensors provide the farmers with real time information that they can utilize to make timely decisions on irrigation, fertilizer application, pest management as well as harvesting. One such application is that the soil moisture sensors can be used to determine the dry regions in a field and only the automated irrigation system can be used to pump water to those areas. This saves on wastage of water and ensures maximum moisture of the crops to grow. Similarly, AI systems can provide recommendations based on the data that these sensors gather to determine when to apply fertilizer or pesticide to improve crop health and reduce the number of chemicals applied.
- **Optimization using AI and Data Analytics**: Crop growth, pest, and nutrient requirements forecasting through AI and Data Analytics: AI algorithms can be implemented to analyze the information obtained by the IoT sensors so as to forecast crop growth, pest and nutrient needs. By combining them, farmers will be able to optimize farming activities at a higher degree of precision. As an example, the AI-powered systems would be capable of computing the optimal fertilization level of specific areas of the field based on the soil and crop requirements. The specified solution is not just healthier to crops but also reduces the overutilization of fertilizers, which is also unhealthy to the environment.

- **Field Monitoring and Intervention Drones**: High resolution cameras and sensors are equipped on drones that are currently being used to survey fields in agriculture. These air crafts can capture images of the ground in an in-depth manner and this will help the farmers know the areas that need to be addressed. The drones can identify such problems as pest infestations, crop diseases, or the areas of bad growth and, thus, farmers can, in a more targeted manner, direct their interventions. This precision reduces the need to spray large areas with pesticides or fertilizer and, hence, leads to a low degradation of the environment. In addition, multispectral imaging technology can also be fitted on drones and this will allow farmers to see the health of the plants in different wavelengths which will provide them with a better opportunity of establishing the status of the crops in addition to making necessary adjustments.
- **Precision Spraying Systems**: Pesticides and fertilizer spraying can be made more efficient in precision agriculture, in which drones and AI-based solutions are utilized together. These systems do not spray an entire field, like it is normally done in the application of chemicals, and instead, only spray the areas of the field that require interventions. This approach will significantly reduce waste of chemicals, and will reduce costs and even the impact that agriculture will have on the environment [9].

8.5.2 Sustainable Use of Resources

- Sustainability is the essence of the impact of Industry 5.0 on the agricultural sector. With the increasing pressure on natural resources and the increasing environmental concerns by the traditional farming practices, there is an urgent need to come up with mechanisms that will ensure that maximum utilization of the resources in existence and minimize waste and environmental degradation. Industry 5.0 ensures sustainable agricultural production through the technologies that enable the production to be more efficient, waste-reducing, as well as resources-saving (water, land, energy).
- **Smart Irrigation Systems**: Water is one of the most valuable resources in the sphere of agriculture and its sustainable usage is the key to its farming. Smart irrigation through IoT sensors can provide the potential to analyze the soil moisture, weather forecasts and crop water requirements and adjust water use dynamically. The irrigation system will be turned off or delayed to water contentious and thereby conserve water and wastage contrary to the case where the sensors detect that the moisture content in the soil is satisfactory, thereby triggering the irrigation system. On the same note, AI systems can study weather patterns to forecast future rain and schedule irrigation to provide crops with the appropriate water at the appropriate time.
- **Robotic Harvesters and Artificial Intelligence Planting Systems**: Robotics is also becoming central in reducing the number of resources used in farming. An example is robotic harvesters, which have the advantage of harvesting fruits and

vegetables at the right time without causing damage to the crops and, hence, wastage. The robots can also work 24 h, enhancing the harvest and lowering labor expenses. Moreover, AI-assisted planting systems make the most out of seed placement by planting crops at the optimal depth and spacing. These systems decrease wastage and maximize productivity because they allow the efficient use of seeds, water, and fertilizers, which make farming more resource efficient.

- **Vertical Farming**: Vertical farming is another sustainable method of agriculture in the scope of Industry 5.0 and is supported by AI and IoT technologies. Vertical farming is the practice of cultivating crops in vertically inclined planes or in layers, which saves much land to be used in production. The technique can be quite beneficial in urban centers where land that can be used to grow crops is limited. Vertical farms are hydroponics or aeroponic, which means that they use nutrient-rich water solutions instead of soil, so they are more water-efficient. Besides, such farms are energy-efficient due to controlled environments, the climate control system that AI controls light, temperature, and humidity to provide ideal growing conditions throughout the year. Consequently, vertical farming is a viable alternative to food production in areas with high population density, thus reducing the impact of agriculture on the environment [10].
- **Case Study**: Aero farms are pioneers of vertical farming that make use of aeroponic systems to cultivate crops in a controlled environment indoors. Vertical farms such as Aero farms allow more sustainable production of food by using much less water and land than conventional agriculture, relieving pressure on natural resources. Also, AI can be used to observe and manipulate the environment to maximize crop growth so that the yields are maximized with minimal environmental costs.

Figure 8.3 illustrates the adoption of a high-tech implementation in the contemporary agricultural industry, in that case, precision farming. A tractor is also working on the field with a robot arm serving the crops, and a drone is looking down on the field. One can see a weather sensor, which underlines the importance of IoT and AI in streamlining the agricultural processes, including irrigation and the use of pesticides. This presents the sustainable resource utilisation and data-driven decision-making through Industry 5.0 in the agricultural industry.

8.5.3 Benefits

The synergies between the idea of precision farming and sustainable use of resources, along with the assistance of Industry 5.0, have several benefits that can transform the agricultural sector and make it competent to solve the problems of global food supply and mitigate its adverse impact on the environment.

- **Enhanced Crop Outputs**: Precision farming optimizes the investor of resources, water, fertilisers and pesticides to boost crop production and also ensure the

Fig. 8.3 Smart farming: data-driven solutions for a sustainable future

production of food is more stable. This kind of efficiency in resource utilization helps farmers to grow more crops in less area of land and, therefore, leads to their prosperity and greater food security.

- **Environmental Impact**: Environment impact is among the highest benefits of precision farming as the effect of the same is significantly minimal on the environment. The use of smart irrigation, AI-controlled fertilizing, and crop monitoring made with the assistance of drones ensures the maximum reduction of wastes, water, and harmful chemicals, and the need to use them is reduced. The practices help in conservation of biodiversity and prevention of soil erosion, and also prevent water quality by preventing the flow of pesticides and fertilizers into other ecosystems.
- **Cost Saving**: Precision farming is cost saving that utilizes fewer resources and makes the most out of the farm operations which involve water, energy and fertilizers. These savings can significantly increase profitability of the farms, especially to the smallholder farmers who would otherwise not be able to meet the rising cost of input. Also, automation and robotics lead to the decline of the labor costs and hence farmers are still profitable due to the rising records of labor shortages and wage inflation.
- **Food Security**: Precision farming is also a significant aspect in enhancing food security in the globe. It improves productivity and efficiency, therefore, more reliable food production, which is highly crucial in the fluctuating weather patterns, population increase and land erosion. The ability to produce food more with less land and less resource makes agriculture more resilient to the environment stress to ensure food production in the future can be carried out sustainably [11].

In conclusion, Industry 5.0 has offered new opportunities to make the agricultural system more efficient, productive, and environmentally friendly due to the new possibilities in precision farming and sustainable resource utilization that can become the game-changer. Using AI, IoT, robotics, and other new technologies will enable farmers to simplify their jobs and reduce wastage as well as ensure the sustainability of food production in the long term. Such technologies are yet to be developed and will continue to provide more potential to make world agriculture more efficient and sustainable.

8.6 Smart Cities: Infrastructure, Energy Management and Engagement of Citizens

Industry 5.0 is greatly contributing to the future of the urban life since it is transforming cities into smart, connected and at the intersection of technology and people interaction. The aim of developing smart cities is to provide the citizens with a better life, be more sustainable, and empower citizens with new advanced technologies, such as the Internet of Things (IoT), artificial intelligence (AI), big data, and robotics. Some of the primary sides to this transformation are the process of accumulating smart infrastructure and smart energy management instruments and citizen involvement. These technologies do not only render the cities more efficient in terms of their management and operations, but also make the city more dynamic and open to the residents.

8.6.1 Smart Infrastructure

The basis of a smart city is smart infrastructure whereby the physical infrastructure is equipped with sensors, AI systems and connectivity to monitor and manage urban systems in real time. Technology provides a chance to use the resources more effectively, preserve them, and adjust to the changes that take place in the city on a dynamic level.

- **Smart Buildings and Roads**: The simplest examples of smart infrastructure are smart buildings and smart roads that are made possible through the aid of IoT. Smart buildings are detectors installed to check the occupancy, air quality, temperature and energy consumption. The artificial intelligence (AI) systems will be able to control lighting, heating, cooling and ventilation based on occupancy and the weather conditions outside which will maximize energy usage and ensure comfort. A case in point is with an heating, ventilation, and air conditioning (HVAC) system in a smart building, it will be possible to program it to reduce the amount of energy used during low-peak periods or when no-one is present in a building, which saves energy and lowers the carbon footprint of a structure.

- **Smart Roads and Traffic Management**: The transport industry has smart roads with IoT sensors and cameras that facilitate the monitoring of traffic, detecting the congestion and prioritizing the traffic lights over the traffic. The AIs process this data and maximize traffic to prevent traffic jams and make the commute time shorter. As a case in point, smart traffic lights may be programmed to respond to the live traffic information to ensure that the traffic flows more efficiently and less fuel is wasted. Moreover, AI systems can predict traffic based on previous information and can offer commuters real-time directions by using such applications as Google Maps or Waze.
- **Digital Twins in Urban Planning**: Urban planning and infrastructures management are among the fields where digital twins, virtual copies of real-life objects are currently being applied substantially. The planners of a city can also do the simulation and testing of various situations by creating digital models of a city or a specific building and using it to apply the changes in reality. Using this as an example, the city planners will be able to simulate different traffic flow systems, the utilization of the urban spaces or the energy consumption trends to come to terms of how all of this will affect the performance of the city. It gives us the opportunity to make knowledgeable decisions, assigning resources in the most optimal manner and contributes to the prevention of costly mistakes. The digital simulations can also predict those changes in the urban environments that might occur in the future such as the increase in the population or even the climatic change, which allows the cities to think about the sustainability over the long term [12].

8.6.2 Energy Management

One of the key concerns of smart cities formation is energy management. The Industry 5.0 technologies are changing the way energy is produced, distributed, and consumed by the cities, and, more specifically, they are preoccupied with the incorporation of renewable energy sources, a greater amount of energy efficiency, and reduced waste levels.

- **AI-Based Energy Optimization**: AI-Based Energy Optimization: this is another AI implementation in the city since AI-based systems are helping to optimize energy consumption in cities by taking real-time data regarding energy use, weather conditions, and energy production. These systems are able to anticipate the energy needs and deliver and maintain a supply of energy made available by renewable sources of energy like solar or wind based on the demand of the city which fluctuates. As an example, AI systems can be able to use solar energy to power buildings, streetlights, and other public transport systems during high solar energy generation periods (e.g. sunny days). On the other hand, the system can pull energy through other renewable sources or the grid when the production of solar energy is low to ensure a uniform supply of energy.

- **Efficient Power Distribution through Smart Grids**: Smart grids are the most necessary component of the distribution of power in a smart city as IoT sensors and AI power them. These grids ensure that the consumption of electricity is monitored in real time, and hence, any form of energy waste can be corrected in real-time to enhance the efficiency of the power distribution system. To illustrate, in the case when a certain segment of the city shows increased energy consumption compared to the estimated numbers, the intelligent grid will be able to re-distribute the energy to the less-demanding locations, maintaining the equilibrium of the supply within the city. Moreover, smart grids can integrate energy storage facilities, like batteries, to accumulate the surplus energy produced at the peak times of renewable energy production and utilize it at the low times.
- **Renewable Energy**: Cities are using more and more renewable energy sources such as solar panels, wind turbines, and hydropower to fulfil their energy demands with the incorporation of IoT and AI. Smart energy management systems will also see to it that these sources are utilized to the best of their abilities and minimize the use of fossil fuels and greenhouse gas emissions. An example is that in certain smart cities, streetlights are solar-powered, and they produce energy during the day and automatically vary the brightness at night according to the traffic and pedestrians in order to conserve energy further [13].

8.6.3 Citizen Engagement

The key element of the smart city's success is citizen engagement because it allows the citizens who live in these cities to be heard in the decisions that are made on their behalf and to be able to take an active role in transforming the urban environment. Industry 5.0 is improving how citizens engage with the local governments and how they consume public services by using digital platforms and AI-driven systems.

- **Digital Platforms of Public Services**: Digital platforms and apps are one of the most direct interactions smart cities have with their citizens. These sites enable citizens to get live information about city services, events and emergencies. The citizens will be able to get notified about the traffic, the schedule of public transport, weather, or emergencies such as fire or floods. Moreover, citizens may directly notify local authorities about problems, like potholes, broken streetlights or water leaks, with the help of those platforms. The data is processed, and issues are prioritized so that they can be solved immediately, thus enhancing the responsiveness and effectiveness of the services of the city.
- **AI Chatbots and Virtual Assistants**: To go even further in the dimension of citizen engagement, AI chatbots and virtual assistants are becoming more common in order to facilitate personal communication between residents and local governments. These chatbots are able to respond to questions, facilitate requests to city services, notify about the status of the problem, and even help citizens to use local government services, like applying for a permit or paying a fine. The resident may enter a chatbot to ask how the request to fix a streetlight that is broken

or what information can be found on recycling services, etc. The latter can be significantly streamlined with the assistance of AI-based virtual assistants that can provide an immediate reply and help streamline the process of delivering the services provided by the government.

- **Feedback on Urban Enhancements**: Smart cities also make use of digital resources to receive real-time feedback of the citizens to ensure that the local government can better understand the needs and preferences of the citizens. Such feedback is founded on data help in urban planning and policymaking that is that the wishes of the people can be included in the development of a city. To give only one example, mobile apps or social media websites could be used to pollute its citizens on what they would prefer in their local areas in the form of green spaces, public transport lines, or facilities. A sense of community and trust is established between the residents and the city authorities through such a participatory process [14].

Figure 8.4 describes a dystopian urban landscape, which has a number of symbols denoting intelligent features of the city. It emphasizes the combination of intelligent infrastructure, energy, and participation of citizen interactions. Wind turbines, solar panels, traffic lights, smart buildings, and mobile devices can be considered visual elements, which is symbolic of how AI, IoT, and big data are used to control the city and the interaction between its citizens. The idea of the digital interaction is presented using a smartphone and chat bubbles as the real-time communication tools.

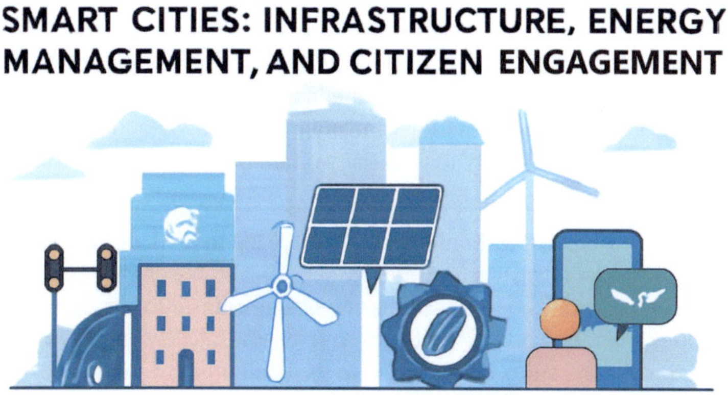

Fig. 8.4 Transforming cities with smart tech and citizen engagement

8.6.4 Benefits

The concept of Industry 5.0 is a prospect to bring numerous advantages to urban areas through creating smart cities and making the urban setting more sustainable, with the quality of living improved and a greater number of citizens being involved.

- **Better Quality of Life**: This is due to the fact that smart cities provide a superior access to services, reduced congestion and cleaner environment to the inhabitants. Through intelligent traffic management, the effectiveness of the transportation systems is enhanced resulting in a reduction in the time of commuting and a reduction in the level of pollution. Further, smart infrastructure including garbage and energy systems also contributes to the cleanliness of the streets and facilitate a more sustainable urban life. An example would be the smart trash can sensor which will be capable of detecting the amount of waste in the bins and alert the waste management teams to empty the trash when the bins fill up thereby making the neighborhoods cleaner.
- **Sustainability**: Smart cities are sustainable compared to the traditional urban centers because there is the optimum use of energy, water, and other resources. These cities are contributing to environmental preservation by ensuring that they have renewable energy sources, good energy utilization and reduced wastage by means of smart systems. Water supply management systems can detect water leaks and automatically open and close valves as an example that will reduce the wasting of water and hopefully save the valuable resources.
- **Citizen Empowerment**: Smart cities allow the citizens to be more active in formulating their city as a result of increased access to the governmental service and services provided in the city, increased transparency, and opportunity to become more active in the process of urban planning. This leads to an increased confidence in the government institutions and civic involvement and creates a sense of ownership and accountability toward the welfare of the city [15].

In conclusion, Industry 5.0 is changing the development of the smart cities by introducing the most recent technologies, which are the IoT, AI, and big data, into the development of urban infrastructure, energy system, and the engagement of citizens in the process. These kinds of innovations are turning cities into efficient, sustainable and so people friendly, and this will improve on the quality of life of the urban residents and make the cities better suited to the challenges of the future.

8.7 Cross-Sector Adoption Rates, Barriers, and Enablers

Industry 5.0 will revolutionize most industries and integrate human ingenuity and collaboration with other technologies, such as artificial intelligence (AI), robotics, and the Internet of Things, and cyber-physical systems. In spite of the fact that the potential of Industry 5.0 is high, its usage within the numerous sectors is rather heterogeneous because of such factors as the perspective of investments, the technological maturity, and regulatory policies, and the geographic location. The mentioned variables can influence the pace of adoption and implementation success because they either ease or complicate the adoption and implementation of Industry 5.0 technologies without any delay.

8.7.1 Cross-Sector Adoption Rates

The implementation of Industry 5.0 is being done in different industries, but its rate and the degree of adoption will depend on the specific industry due to the level of technological maturity, regulation, and other economic considerations.

- **Manufacturing**: Manufacturing is one of the industries that have been most sophisticated in terms of embracing the 5.0 industry technologies. The use of such technologies as collaborative robots (cobots), AI-based production lines, and digital twins has already been incorporated by many large enterprises to maximize production efficiency and enhance customization possibilities. These technologies facilitate faster manufacturing processes and customized products as they are ordered. Nevertheless, the rate of adoption is usually lower among small and medium-sized enterprises (SMEs) because of the initial expenses involved in the acquisition and installation of these superior technologies. SMEs are not usually able to afford such innovations or even have the technical skills to implement such innovations on a large scale, and therefore, they are slower to adopt such innovations in this segment of the manufacturing sector. Filling this gap is the issue which is pivotal to the wider adoption in the sector.
- **Healthcare**: The healthcare industry has also been gradually implementing Industry 5.0 technologies, specifically in the fields of robotic surgery, individualized medicine, and systems of monitoring patients. Although the largest hospitals and research facilities are at the forefront, the implementation is less liberal in smaller clinics and in rural healthcare networks. Regulatory barriers, the cost, and the difficulty of incorporating new technologies within the current healthcare systems are the factors that slow down progress. Also, patient data privacy concerns and health technology regulatory aspects might bring an additional layer to the implementation process and make it a more gradual and cautious procedure as compared to other industries.
- **Agriculture**: The farming business is rapidly embracing the Industry 5.0 technology, particularly in the form of precision farming, IoT-based irrigation, and AI-based crop management service. By using this kind of technologies, farmers are given a chance to optimize the operating resources like water, fertilizers, and energy and this increases productivity and sustainability. But its mass application, particularly in the rural sphere or in poorly developed regions, is inhibited by low infrastructure, high investment demands and unavailability of technical support. Also, smallholder farmers might not be able to embrace the use of highly

advanced technologies because they do not have sufficient access to capital, which is a further impediment to the implementation of precision farming solutions on a large scale.

- **Smart Cities**: Most cities globally have started to use smart infrastructure, energy systems, and IoT-based services to enhance the quality of life in cities. The purpose of these technologies is to optimize the traffic, improve energy consumption and have sustainable urban development. Nevertheless, the implementation of smart city technologies is not universal, and megacities are usually at the forefront, whereas small cities or cities in developing nations are challenged by the lack of finances, infrastructure, and legacy systems. Also, the interoperability of various technologies and systems of various vendors poses a challenge and scaling, and long-term sustainability of smart cities is a complicated process [16].

8.7.2 Barriers

Industry 5.0 technology is barred by a number of obstacles in different fields. These challenges may delay the process of digital transformation and hinder the ability of industries to use the potential benefits of Industry 5.0 to the full extent.

- **Cost and Investment**: The high initial investment costs required to acquire and implement new technologies are one of the main obstacles to the adoption of Industry 5.0. Most of the innovative solutions, including robotic systems, AI infrastructure, and IoT networks, demand high initial investments. Small and medium-sized businesses (SMEs) or even businesses in developing economies may not have the funds or the means to invest in such technologies. Implementation cost frequently consists not only of the price of the technology itself but also of the price of training employees, incorporating the systems into existing infrastructure, and long-term maintenance. The rate of adoption may be slow without enough financial support or government incentives.
- **Technological Complexity**: This is another highly significant challenge the complexity of integrating the 5.0 technology industry into the legacy systems. The existing infrastructure may not be compatible with the new advanced digital technologies in the manufacturing industries as well as healthcare and may not integrate readily. This is particularly true in the industries where the systems and processes are mature and where a large portion of reengineering and retraining is involved in implementing the new technology in the systems. As an example of the healthcare industry, the use of AI-based diagnostic technologies or robotic surgery systems is often associated with the process of integrating them into the already existing medical equipment and electronic health records (EHR) systems and addressing regulatory demands. Considering the same, manufacturing can also imply a redesign of the traditional production lines to include collaborative robots and AI-based production lines, which can be both a time-consuming and an expensive procedure.

- **Regulatory and Compliance Issues**: The regulatory and compliance problems of these industries as healthcare and smart cities are rather strict and slow down the introduction of technologies of Industry 5.0. One example in the field of healthcare is that patient privacy and regulatory laws such as the Health Insurance Portability and Accountability Act (HIPAA) in the U.S. or General Data Protection Regulation (GDPR) in the EU are a significant obstacle towards the application of AI-driven solutions. Similarly, the mechanism of adoption can be delayed in smart cities due to the necessity to make sure that the local, national, and international regulations are adhered to in data protection, cybersecurity, and urban planning. These regulatory environments and the problem of data security and privacy are complicated, and that is why it is difficult to implement new technologies on a large scale in a short period of time [17].

8.7.3 Enablers

Nevertheless, despite the challenges, there are several aspects that make the rapid adoption of Industry 5.0 technologies in most industries easy. These facilitators act in the opposite direction and ease the process of moving towards more advanced technological settings.

- **Government Support**: Governments will play an essential role in accelerating the adoption of Industry 5.0 technologies. Policy programs, grant funding, government-corporate collaborations, and infrastructure projects are the key facilitating variables. Governments of most countries have incentives to encourage businesses to adopt new technology including tax breaks, subsidies and grants. Besides, governments play the role in creating favorable administrative environments that present transparency to companies that are interested in implementing creative solutions. To illustrate this point, in the European Union, the use of digital technology is being encouraged by such initiatives like Digital Single Market, which aim to harmonize the regulations and give SMEs a chance to obtain funding. The government also supports public infrastructure developments such as smart city development which can be expensive to venture into by the municipalities on their own.
- **Technological developments**: The long-term technological advancements in the AI, IoT, robotics and data mining areas are democratizing the industry 5.0 technologies and rendering them affordable. As these technologies continue to develop, they are becoming more powerful, and their prices are becoming lower, so they can be used even by small-sized businesses. To illustrate this, the increasing popularity of AI platforms that are based on clouds and IoT networks allows even small businesses to experience the utility of highly efficient analytics tools without necessarily making significant investments in the physical infrastructure. It also reduces the level of technological complexity of implementation because the development of user-friendly interfaces and integrated solutions does not require the use of experts to manage and use such systems.

- **Public Awareness and Consumer Demand**: With the population being increasingly aware of the benefits of Industry 5.0, consumers are increasingly generating demand of individualized products, smart service and green solutions. This consumerism compels companies to merge them with the new technology in order to stay competitive and meet the emerging needs. As an example, the growth of personalized medicine and sustainable products makes healthcare delivery, producers, and retailers ready to invest in AI and IoT devices that can offer customized assistance. Another driver behind smart technologies is the increasing demands of citizens in smart cities to have better urban services, networks and transportation, and a clean environment [18].

8.8 Learning Outcomes of the Chapter

Upon completing this chapter, readers will be able to:

- Understand the fundamental concepts and applications of Industry 5.0 technologies in key sectors such as manufacturing, healthcare, agriculture, and smart cities.
- Identify the challenges and benefits associated with the implementation of Industry 5.0 technologies in these sectors.
- Evaluate how Industry 5.0 fosters collaboration between humans and machines, leading to enhanced customization, operational efficiency, and sustainability.
- Analyze the barriers to the adoption of Industry 5.0 technologies, such as cost, complexity, and regulatory issues, and understand the strategies to overcome these challenges.
- Recognize the enablers, such as government support, technological advancements, and consumer demand, which facilitate the widespread integration of Industry 5.0.
- Apply knowledge of Industry 5.0 to real-world case studies and examples in manufacturing, healthcare, agriculture, and smart cities.

8.9 Review Questions of the Chapter

1. How does Industry 5.0 shift the traditional manufacturing process towards customization and collaborative production? Discuss the role of cobots, AI, and digital twins in optimizing production lines for tailored products.
2. In what ways does Industry 5.0 enhance healthcare through personalized medicine, robotic surgery, and real-time patient monitoring? Provide specific examples of how these technologies improve patient outcomes and reduce healthcare costs.
3. Discuss the impact of precision farming on resource conservation and crop yield optimization. How do IoT sensors, AI, and drones contribute to sustainable farming practices under Industry 5.0?

4. How can smart city technologies such as digital twins, IoT-based energy management systems, and citizen engagement platforms improve urban living? Analyze how these technologies contribute to sustainability and quality of life in smart cities.
5. What are the key challenges that sectors like healthcare and agriculture face in adopting Industry 5.0 technologies? Discuss the regulatory, infrastructural, and financial barriers to widespread adoption.
6. Evaluate the role of AI and IoT in creating smarter, more adaptable manufacturing environments. How does real-time data monitoring and predictive analytics contribute to improving production processes and reducing waste?
7. Industry 5.0 emphasizes human–machine collaboration. How does this human-centric approach differ from previous industrial revolutions, and what are the benefits and challenges of integrating humans with advanced technologies in sectors like manufacturing and healthcare?
8. Discuss how the concept of mass customization in manufacturing is facilitated by Industry 5.0 technologies. How do cobots and 3D printing enable manufacturers to produce personalized products at scale without sacrificing efficiency?
9. What role does government support and public–private partnerships play in overcoming the barriers to Industry 5.0 adoption? Provide examples of how policy initiatives can accelerate the implementation of these technologies in sectors like agriculture and smart cities.
10. Assess the potential for Industry 5.0 to drive economic growth and job creation in sectors such as manufacturing and healthcare. How does the shift towards human-centric, technology-driven industries create new employment opportunities?

Appendix: Solution Manual

1. **How does Industry 5.0 shift the traditional manufacturing process towards customization and collaborative production? Discuss the role of cobots, AI, and digital twins in optimizing production lines for tailored products.**

   Solution: Industry 5.0 shifts traditional manufacturing by incorporating cobots (collaborative robots) and AI into production lines, enabling mass customization. Cobots work alongside human operators, handling repetitive or hazardous tasks while humans focus on creativity and strategic decision-making. Digital twins, by providing virtual replicas of physical production lines, allow for real-time adjustments, optimizing workflows and reducing waste while maintaining personalized production.

2. **In what ways does Industry 5.0 enhance healthcare through personalized medicine, robotic surgery, and real-time patient monitoring? Provide specific examples of how these technologies improve patient outcomes and reduce healthcare costs.**

   Solution: Industry 5.0 enhances healthcare through AI-driven personalized medicine, where algorithms analyze genetic data to recommend tailored treatments. Robotic surgery systems, like the da Vinci system, enable minimally invasive procedures with high precision, reducing recovery times. IoT-based patient monitoring continuously tracks vital signs, enabling early intervention and reducing the need for hospitalization, thus cutting costs and improving patient outcomes.
3. **Discuss the impact of precision farming on resource conservation and crop yield optimization. How do IoT sensors, AI, and drones contribute to sustainable farming practices under Industry 5.0?**

   Solution: Precision farming in Industry 5.0 leverages IoT sensors to monitor soil health, weather conditions, and crop status. AI analyzes this data to optimize resource use—adjusting irrigation schedules, applying fertilizers, and controlling pests with minimal environmental impact. Drones help with field monitoring, allowing for targeted interventions that boost crop yield while conserving water and reducing the need for chemical inputs.
4. **How can smart city technologies such as digital twins, IoT-based energy management systems, and citizen engagement platforms improve urban living? Analyze how these technologies contribute to sustainability and quality of life in smart cities.**

   Solution: Smart city technologies like digital twins model urban infrastructure, allowing city planners to simulate traffic flow, energy usage, and space utilization. IoT-based energy management systems optimize power distribution, reduce waste, and ensure sustainability. Citizen engagement platforms, through apps and AI chatbots, improve communication with local governments, enabling real-time updates and enhancing trust and participation in city governance.
5. **What are the key challenges that sectors like healthcare and agriculture face in adopting Industry 5.0 technologies? Discuss the regulatory, infrastructural, and financial barriers to widespread adoption.**

   Solution: Healthcare and agriculture sectors face challenges such as high initial costs, complex regulatory requirements, and infrastructure limitations. Healthcare must adhere to strict privacy regulations (e.g., HIPAA), which can slow down the adoption of IoT and AI technologies. In agriculture, the lack of digital infrastructure in rural areas and the high cost of precision farming technologies can hinder widespread adoption, despite the environmental and economic benefits.

6. **Evaluate the role of AI and IoT in creating smarter, more adaptable manufacturing environments. How does real-time data monitoring and predictive analytics contribute to improving production processes and reducing waste?**

   Solution: AI and IoT allow for real-time data collection and analysis, enabling predictive maintenance and process optimization in manufacturing. Sensors and digital twins provide data on equipment health, predicting failures before they occur, thereby reducing downtime. AI-driven algorithms can also optimize production schedules, minimize waste, and ensure that resources are allocated efficiently, improving overall productivity.
7. **Industry 5.0 emphasizes human–machine collaboration. How does this human-centric approach differ from previous industrial revolutions, and what are the benefits and challenges of integrating humans with advanced technologies in sectors like manufacturing and healthcare?**

   Solution: Unlike previous industrial revolutions that emphasized automation and efficiency at the expense of human roles, Industry 5.0 emphasizes collaboration between humans and machines. Cobots, AI, and other technologies work alongside workers, enhancing their capabilities rather than replacing them. The challenge lies in ensuring that workers have the necessary skills to collaborate with advanced technologies, but the benefit is a more engaged, skilled workforce focused on higher-level decision-making and creativity.
8. **Discuss how the concept of mass customization in manufacturing is facilitated by Industry 5.0 technologies. How do cobots and 3D printing enable manufacturers to produce personalized products at scale without sacrificing efficiency?**

   Solution: Industry 5.0 enables mass customization by combining AI, cobots, and 3D printing. Cobots help create flexible production lines that adapt quickly to new product designs, while 3D printing allows for tailored products at scale. These technologies reduce the cost of customization by minimizing waste, allowing manufacturers to produce unique items without sacrificing efficiency.
9. **What role does government support and public–private partnerships play in overcoming the barriers to Industry 5.0 adoption? Provide examples of how policy initiatives can accelerate the implementation of these technologies in sectors like agriculture and smart cities.**

   Solution: Government support, through subsidies, tax incentives, and funding for research and development, can help overcome the financial barriers to Industry 5.0 adoption. Public–private partnerships can foster collaboration between industries, governments, and educational institutions to build infrastructure, enhance workforce skills, and develop technologies that drive innovation in sectors like agriculture and smart cities.

10. **Assess the potential for Industry 5.0 to drive economic growth and job creation in sectors such as manufacturing and healthcare. How does the shift towards human-centric, technology-driven industries create new employment opportunities?**

    Solution: Industry 5.0 creates opportunities for new roles in areas like AI management, robotics programming, and data analysis. By focusing on human creativity and collaboration, industries can shift away from purely manual labor to more strategic, innovation-driven roles. These new jobs will drive economic growth by creating a more skilled and adaptable workforce, benefiting sectors like manufacturing and healthcare.

Bibliography

1. Shabur MA, Shahriar A, Ara MA (2025) From automation to collaboration: exploring the impact of industry 5.0 on sustainable manufacturing. Discov Sustain 6(1):341
2. Marinelli M (2023) From industry 4.0 to construction 5.0: exploring the path towards human–robot collaboration in construction. Systems 11(3):152
3. Jwo J-S, Lin C-S, Lee C-H (2021) Smart technology—driven aspects for human-in-the-loop smart manufacturing. Int J Adv Manufact Technol 114(5):1741–1752
4. Khan WA et al (2025) Industry 5.0 computer aided engineering design and manufacturing: a fourth industrial revolution perspective. Springer, Cham, pp 523–535
5. Basulo-Ribeiro J, Teixeira L (2024) The future of healthcare with industry 5.0: preliminary interview-based qualitative analysis. Fut Internet 16(3):68
6. Mukherjee A et al (2024) Cobots: enhancing human creativity and productivity in Industry 5.0. In: Powering Industry 5.0 and sustainable development through innovation. IGI Global, pp 186–197
7. Hammad M et al (2023) Framework for the implementation of smart manufacturing systems: a case in point. Processes 11(5):1436
8. Adel A, Alani NHS (2024) Human-centric collaboration and Industry 5.0 framework in smart cities and communities: fostering sustainable development goals 3, 4, 9, and 11 in society 5.0. Smart Cit 7(4):1723–1775
9. Singh G, Sharma S (2025) A comprehensive review on the internet of things in precision agriculture. Multimedia Tools Appl 84(17):18123–18198
10. Alreshidi E (2019) Smart sustainable agriculture (SSA) solution underpinned by internet of things (IoT) and artificial intelligence (AI). arXiv preprint arXiv:1906.03106
11. Zhang Y (2024) A path to sustainable development of agri-industries: analysis of agriculture 5.0 versus industry 5.0 using stakeholder theory with moderation of environmental policy. Sustain Develop 32(5):4829–4843
12. Adel A (2025) Framework for human-centric smart cities integrating Industry 5.0
13. Ramírez-Gordillo T et al (2023) Human-centric solutions and AI in the smart city context: the industry 5.0 perspective. In: The international research and innovation forum. Springer, Cham
14. Hammad M et al (2025) Transforming the electronics manufacturing sector. In: Smart manufacturing blueprint: navigating industry 4.0 across diverse sectors. Springer, Cham, pp 115–163
15. Lopez R, Juan L, Castro AIG (2020) Sustainability and resilience in smart city planning: a review. Sustainability 13(1):181
16. Movahed NA, Nozari H, Movahed AB (2025) Systematic review of smart factories production in industry 5.0

17. Hammad M et al (2025) Smart manufacturing in food and beverage industry. In: Smart manufacturing blueprint: navigating industry 4.0 across diverse sectors. Springer, Cham, pp 213–243
18. Ashley PA (2010) Interactions between states and markets in a global context of change: contribution for building a research agenda on stakeholders' social responsibility. Int Instit Soc Stud 506:53232

Part VI
Adoption Barriers, Case Studies, and Future Outlook

Chapter 9
Embracing Industry 5.0: Overcoming Challenges and Leveraging Key Strategies

9.1 Objectives

- To determine and discuss the major technological, financial, cultural, and regulatory obstacles to the implementation of Industry 5.0 technologies.
- To study the issues related to the implementation of such advanced technologies as AI, robotics, IoT, and digital twins into the current systems.
- To examine the financial barrier of Industry 5.0 and SMEs, especially SMEs, to adopt the idea and the method of ROI analysis.
- To talk about the cultural opposition to automation and how the change management strategies can take care of the workforce issues.
- To assess the contribution of policy and regulatory frameworks to the promotion or restriction of the adoption of the technologies of Industry 5.0 on the scale.
- To recommend strategic models, including SWOT analysis, to be used in breaking adoption barriers and successful integration of Industry 5.0 in different sectors.

9.2 Overview

Industry 5.0 is the following stage of industrial development, where human intelligence, highly developed machines, and automation are on their way to merging. As opposed to its predecessor, Industry 4.0, which focused on the full automation of processes, Industry 5.0 incorporates the role of humans and their creativity and decision-making in the same scope as robots and AI. The driving factor behind this change is the technology that allows it, the need for society to have personalized products, and the pressure to follow the sustainability agenda.

Industry 5.0 is all about a human being in the center of the industrial processes. It is a revolution in which machines and AI are not meant to do the work of humans but to amplify their abilities. This part will describe the fundamental principles of

M. Hammad et al., *Industry 5.0: The Human-Centric Future of Industry*, Springer Series in Advanced Manufacturing, https://doi.org/10.1007/978-3-032-12622-1_9

Industry 5.0, such as the interaction of humans and machines, intelligent automation, and sustainability. The major attributes like personalization, flexibility, and ethical production will also be addressed.

9.3 The Need for Industry 5.0

The rising demand for Industry 5.0 is based on some of the most important factors that are changing the global industrial environment and growing attention to sustainability, ethical aspects, and human well-being. Although Industry 4.0 has introduced unprecedented levels of automation and efficiency, it has also given rise to certain problems that should be overcome so that technological development would not contradict the values and interests of society [1]. These difficulties are some of the followings:

- **Job Losses and Human Displacement**: This is one of the criticisms under Industry 4.0 since the industry is likely to cause job losses when machines replace human labor. Industry 5.0 will solve this problem by first focusing on human participation in the process and making sure that the machines will not substitute for human workers but instead will supplement them. According to this model, there is a need to strike a balance between technology and human input where the workers will be able to perform higher-level functions like creativity, innovation, and making complex decisions.
- **Non-personalization**: The technologies of Industry 4.0 are currently being optimized to work efficiently and be able to produce large masses of goods, which is why they are not always able to serve the individual needs of customers and take personalization into account. This is the gap that Industry 5.0 is filling because it promotes innovative solutions where human employees will be able to employ their imagination to create products and services that will fit the personal taste of the customers and, therefore, increase their satisfaction.
- **Sustainability and Environmental Responsibility**: With increasing automation in industries, mass production has become an issue of concern to the environment. Industry 5.0 proposes the sustainability of production, which involves AI and IoT to make the best use of energy, minimize wastage, and use fewer resources. It also encourages the use of renewable energy and materials that are environment friendly in the process of carrying out manufacturing processes such that the industrial development does not cut across with environmental conservation.
- **Ethical Concerns**: The implementation of AI and automation has introduced such ethical issues as data privacy, biases of the algorithm, the openness of the automation-driven decisions, etc. Industry 5.0 also lays a lot of emphasis on ethical AI and responsibility and ensures that high-tech technologies are utilized in a responsible manner. This entails establishing clear rules and structures to manage ethical use of technology by means of fairness, transparency and observing human rights.

- **Workforce Well-Being**: The health concern about the well-being and mental condition of the staff and the more prominent focus on the latter is another significant driver behind Industry 5.0. Burnout, stress and dissatisfaction of the employees may also be the result of high demands and monotony of work that are the new standards in the modern production environment. Industry 5.0 promotes a more humanistic workplace where the physical and psychological health of the workers should be put in the first place. Industry 5.0 helps to make the working process more rewarding, making people more productive and satisfied with their job and the atmosphere of innovation and cooperation between humans and machines is established [2, 3].

This Fig. 9.1 conceptually depicts the notion of Industry 5.0, with the focus on the balance between human employees and high automation. The humanoid robot and the worker are shown as shaking hands which symbolizes cooperation and respect. On top of them there are the icons that symbolize some of the main components of Industry 5.0: sustainability (a leaf), creativity (a brain), ethics (scales of justice), and the well-being of the workforce (a human figure). Its background adds a factory environment that has automated equipment, and a modern city skyline, which further supports the use of technology in an urbanized world.

Fig. 9.1 A new era of innovation: Industry 5.0 focuses on ethics, well-being, and environmental responsibility

9.4 Key Strategies for Transitioning to Industry 5.0

With the transition of industries to Industry 5.0, which implies the implementation of human-friendly technologies, collaborative innovation, and sustainable approaches, long-term success requires acceptance. The following are the most important strategies that can be embraced by industries in order to enable such a transition and to ensure that human and technological resources can be optimized in the best possible in terms of efficiency, creativity, and sustainability [4].

i. **Human–Machine Collaboration**

- **Integration of Human Creativity**: In Industry 5.0, the idea that machines will substitute the human workforce is replaced with the idea that machines will enhance human capacity. Whereas Industry 4.0 was all about automation of tasks and processes, the new Industry 5.0 presents a collaborative setting where humans and machines work together and make use of their strengths. AI and robotics are able to take the task of repetition or complex tasks and do them with precision, but it is human creativity, problem-solving, and judgment that are the keys to innovation and unique solutions.

  Industries are advised to invest in technologies that can help with human creativity, such as design tools with the aid of AI or collaborative robots, also known as cobots, which give humans superior abilities. As an example, in the design of cars, AI can be used to model various designs, and designers concentrate on design customization to the demands of the market to ensure product innovation and customization.
- **Personalized Automation**: Industry 5.0 focuses on the idea of personalized automation, i.e., systems that are flexible enough to suit the requirements of individual workers, not to substitute them. In manufacturing, cobots may be applied to help employees by doing heavy lifting, automating repetitive tasks, and increasing the efficiency of operations so that human employees can make more strategic, creative, and complex decision-making for them. As an example, in assembly lines, cobots might perform such tasks as material handling or quality control, and human workers will be involved in troubleshooting, adjustments of product design, and customer-related changes.

  This collaboration of human creativity with automation will make sure that the industries not only enhance their efficiency but also will be able to add personalized value to their products and services, which will lead to customer satisfaction and innovation [5].

ii. **Upskilling and Reskilling Workforce**

- **Training in High Technological Skills**: Because of the implementation of high technologies in industries such as AI, IoT, and cobots, it is important to make sure that the workforce is ready to deal with the changes. There should be training programs that are human-centered, like collaborative robotics, AI interaction, and sensor-based technologies.

Training on upskilling should provide employees with skills on how to work, maintain, and cooperate with advanced technology. As an example, the workforce will be ready to take over the jobs created in Industry 5.0 as a result of training in the ways of programming and interacting with robots, using the power of AI-driven intelligence, or utilizing IoT sensors to observe production processes. Moreover, ensuring that the workers are able to adjust to the technological changes, the workers will also be part and parcel in the success of their organizations.

- **Soft Skills Development**: In spite of the fact that technical skills are significant, Industry 5.0 is also closely concerned with soft skills, which are also considered to play a critical role in the course of interaction between a human and a machine. They should possess critical thinking, creative, problem-solving, leadership and emotion intelligent workers. The soft skills enable the workers to have a major role in the decision making process that involves both machines and human beings.

  Organizations need to make soft skills training part of the program and encourage employees to learn good interpersonal communication skills, teamwork and problem solving skills. Through this development, the industries will ensure that workers fit in an environment that is collaborative and highly technological and increase their capability to pioneer innovative projects.

iii. **Sustainability and Ethics Integration**

- **Green and Sustainable Technologies**: Sustainable 5.0 has taken sustainability as one of its priorities. The industry must change to green technologies, which will be less harmful to the environment, such as using renewable energy-based equipment and closed production. AI and IoT technology can do a lot in the area of resources optimization, waste reduction, and energy efficiency.
- An example is that the production process can be monitored in real-time with the help of AI to identify inefficiencies or wastage. Industries that are environmentally friendly will be implemented by incorporating sustainable materials and energy efficient technologies into the industries to ensure that they are not only productive but also environmentally friendly. Among the circular economic processes that ought to be adopted to ensure that industrial operations have minimal ecological footprint, recycling and reuse of waste materials should be recognized.
- **Ethical AI Use**: As the application of the AI in the industrial practice is increasing, the most significant is ethical concerns related to the usage of AI. The application of AI and robotics is to be transparent, equitable, and open to accountability to avoid the loss of such issues as algorithmic bias, data privacy, and unfair labor practices.

It is necessary to set specific rules and ethical systems for using AI. As an example, the industries should make sure that the data on which the AI models are trained is diverse and representative to avoid bias. Moreover, data privacy policies should be stringently followed as well so that the personal information utilized by AI systems should not be misused. Industries must also take a pledge to make AI deployment equitable, such that automation only boosts human ability without affecting fairness and employment possibilities.

iv. **The Future of Flexible and Agile Manufacturing Ecosystem**

- **Production Systems that can be Tailored**: The fifth generation of the industry is associated with the possibility of responding to the changes in demand and market in the shortest period. Customizable production systems are also flexible and easy to reprogram or reconfigure, unlike conventional systems that are rigid and inflexible and can only be reprogrammed or reconfigured with difficulty.

  By investing in flexible manufacturing systems that are based on modular machines or AI-controlled adjustments, industries can easily scale up production or modify products within a short time without having to face huge losses in terms of time or extra expenses. Such flexibility allows firms to be competitive through the provision of products that meet the customer's needs and, at the same time, ensure efficiency within the company.
- **Smart Supply Chains**: The second important strategy in Industry 5.0 is the use of AI and IoT to develop smart chain supply. The supply chains are able to optimize themselves, and therefore, they make real-time changes in inventory, production, and demand forecasts. With the help of AI and monitoring of the supply chain data in real time, companies can reduce disruptions, predict possible bottlenecks, and guarantee the transparency of the whole supply chain.

  As an example, predictive analytics may be applied to predict inventory shortages, and IoT sensors may help to get current information about stock, the state of the products, and transportation processes. This makes the process more efficient, minimizes wastage, and makes a supply chain more resilient.

v. **Technological Infrastructure**

- **Collaborative Robots (Cobots)**: Industry 5.0 is based on cobots or collaborative robots. Such robots are made to collaborate with human labor, relieving them of physically demanding and monotonous work and letting humans do more complex, creative, and value-related work. The introduction of cobots to the work environment enhances the efficiency and safety of employees since they are not subjected to risky or ergonomically demanding activities.

Integrating cobots allows industries to establish a new working environment where man and machine can work in harmony, and this will increase the level of productivity of the workers and the overall efficiency of operations.

- **Edge Computing and Cloud Technologies**: Edge computing, used together with cloud technologies, should also be implemented by industries to support real-time decision-making, which is necessary for Industry 5.0. Edge computing delays less time in the processing of data by storing and computing it locally on machines or devices, thus making decisions more quickly. Industries are able to gain real-time responsiveness and scale their data storage and management when coupled with cloud computing, which centralizes data storage and management.

  Such two infrastructures facilitate quicker processing of information, less time spent on outages, and the capability of businesses to make decisions based on the latest information available and yet have the capacity to perform large-scale data analysis at the cloud level [6].

vi. **Integrative and Flexible Organization Culture**

- **Creating a Culture of Innovation and Creativity**: Since Industry 5.0 focuses on human creativity, businesses should develop a culture of innovation. Employees should be motivated to experiment, take risks, and find new technologies to promote life-long improvement. Cross-functional teams and innovation labs are great methods to facilitate innovation in the organization.

  Industries could develop an environment where new solutions are generated on a constant basis and improve their competitiveness by creating an environment where the employees are free to think outside the box and bring their new ideas to the table.
- **Employee Well-Being**: The well-being of employees is one of the pillars of Industry 5.0. As technology is becoming more and more connected to the human labor force, there is a need to make sure that the mental, emotional, and physical health of the employees is taken care of by the organization. Wellness programs can be implemented, and this would enable balancing of work and life, stress and mental health resources and thereby an engaged and a productive work force.

  The human-centered technology also implies that the employees will be more ready to feel that their efforts are valued and significant directly resulting in job satisfaction and the overall success of the organization.

vii. **The Cooperation with the Research and Development (R&D) Centers**

- **Partnerships**: It should collaborate with various bodies such as R&D centers, universities and startups to ensure that the company stays on the forefront in terms of technological innovations. These partnerships enable industries to think about new technologies, including AI-related discoveries, human–robot interaction, and ethical AI. Research and development can help industries to come up with solutions to their needs whether it is manufacturing or healthcare which are custom-made.

- **Industry-Specific Solutions**: Not only is it important to have general technological advancements, but the industry should also consider industry-specific advancements, which would deal with industry-specific problems. It may be the design of customized robotic setups to aid in assembling cars to the development of AI-powered diagnostic applications that are likely to benefit the healthcare industry; the R&D alliances can help industries streamline their solutions and be more efficient.

viii. **Regulatory Support and Policy Support**

- **Regulations in the Industry 5.0**: There are a variety of rules that have to evolve with the new technologies and practices introduced by Industry 5.0 so that the process of implementation can be safe, ethical, and sustainable. The policies regarding human–robot cooperation, sustainability, and the application of AI should be changed. The governments should collaborate with the industries to establish systems that will contribute to the responsible adoption of Industry 5.0 technologies [7].

The Fig. 9.2 provides an isometric diagram which graphically sums up the major strategies that industries can pursue to move to Industry 5.0. The plans are projected in the form of 3D icons in a clean and clean manner. It consists of human–machine co-operation, employee skill development, sustainability, responsible AI application, and adaptable production systems. The color palette is well-balanced, and the icons are clear, which demonstrate the idea of Industry 5.0 as human-oriented technologies, collaboration, and sustainability.

Fig. 9.2 Key strategies for transitioning to Industry 5.0

9.5 Challenges and Barriers to Adoption

The development of Industry 5.0 is quite promising, yet it can be characterized by numerous challenges that need to be performed to ensure the spread of technologies. Such barriers fall within the technological, financial, cultural, political, and regulatory barriers. It is particularly crucial in the context of its endeavors to introduce such innovative progressive technologies as AI, robotics, IoT and digital twins to the everyday industrial process. The key barriers and their impact on the implementation of the Industry 5.0 are addressed in this section, followed by a strategic plan to overcome the barriers [8].

i. **Technological Constraints: Integration Complexity and Interoperability Issues**

The integration of the advanced systems with the existing infrastructure is one of the key obstacles to the penetration of Industry 5.0 technologies. A lot of companies are still using legacy systems and have not yet entered the realm of the latest technologies, which are the trademarks of Industry 5.0. Such technological constraints are a great challenge to smooth integration and interoperability.

- **Integration Complexity**
  - **Legacy Systems**: Some of the industries still relying on legacy systems are in the manufacturing and healthcare industries, where the outdated machinery and software are not easily accessible over the latest technologies including AI, robotics, and IoT sensors. The modernization of such systems is a very time- and resource-consuming process that may slow down the process of progressing to Industry 5.0.
  - **Complex System Integration**: The integration of new technologies in workflows and processes may be technically complicated even when the newer technologies have been adopted. E.g., to implement AI and machine learning into a production line, it might be necessary to redesign the workflows, train the personnel, and change the hardware to one that can handle real-time data collection and decision-making.
- **Interoperability Issues**
  - **Non-standardization**: Technology in Industry 5.0 tends to be based on many different platforms, software, and protocols, and this may lead to problems in interoperability. As an example, the IoT devices manufactured by various companies might fail to support effective communication with each other, and the AI algorithms might not be capable of operating in various systems.
  - **Data Silos**: Data silos are still a problem in most industries because data is stored in different departments or systems, and it is not easy to have an overview of operations. With Industry 5.0 to be unlocked, businesses will need to establish data flows without any interruptions and allow data exchange across systems.

- **Solution**
  - **Standardization**: The leaders of the industry must collaborate to develop and implement the universal standards and protocols that would enable smooth integration of the devices and systems of various manufacturers and platforms. To make sure that devices produced by different vendors can effectively communicate with each other, open standards, e.g., IoT protocols (e.g., MQTT, CoAP), should be used.
  - **Implementation of Modular Technology**: Companies are advised to implement modular technology that they can easily incorporate into the existing infrastructure. The creation of edge computing can support the ability to process data locally prior to transmitting it to the cloud systems to simplify the relationship between the older systems and the new technologies.
  - **Incremental Integration**: The companies should not be tempted to make sweeping changes in the systems; the organizations should learn to come up with gradual approaches of implementation where small pilot projects are launched, and performance measured before extending the program.
- **Advantages of the Breaking of Technological Barriers**
  - **Improved Efficiency**: The successful integration of new technologies and old systems can lead to the lack of hiccups in the workflow, failures, and higher productivity of the production.
  - **Quicker Decision-Making**: Interoperable systems may lead to the faster decision-making based on the data and increase the overall level of responsiveness towards the marketplace requirements and the issues of operation.

ii. **Financial Obstacles: Investment and Return on Investment Uncertainties**

The affordability of setting the 5.0 industry technologies could be highly threatening to majority of the organizations and more so to the small and medium companies (SMEs) that could lack the necessary capital to invest in the new advanced systems.

- **Investment Costs**
  - **Expensive startup**: Initiating Industry 5.0 technologies such as cobots, AI-based systems, and IoT infrastructure is costly (both hardware and software-wise). To use a fleet of cobots in a manufacturing facility to use as an example, one will need to factor in the new safety systems, robots, and maintenance on top of the robots.
  - **Long-Term Committals**: The implementation of Industry 5.0 may require the companies to invest on the long-term basis with respect to both the infrastructure and the human resources development and such investment can be very risky of finances and may not be profitable in terms of ROI either.
- **ROI Uncertainties**
  - **Uncertain Returns**: The possible ROI of Industry 5.0, e.g. increased efficiency, improved quality, and reduced downtimes, is self-evident, but it is usually

difficult to estimate, at least at the early stages. Businesses may not be able to estimate the long-term financial impacts of new technology particularly where the benefit is not financial such as customer satisfaction or higher employee productivity [9].
  - **Market Uncertainty**: Industry 5.0 is rapidly changing in terms of technology and so, business may fear to invest in a technology which may be overtaken in the near future, or the business may have to invest more on newer technology.

- **Solution**
  - **Government Incentives**: Government Incentives: Government may provide financial assistance, grants or tax incentives to those businesses that utilize sustainable technologies or automate or invest with AI. This will reduce the initial expense and motivate more industries especially in the developing economies or SMEs.
  - **Public–Private Partnerships (PPP)**: The formation of partnerships between the business and the state may also help to diversify the investment risk and ensure that the companies will afford to implement Industry 5.0 technologies without fully bearing the cost of the investment. Among them is the fact that the local governments can collaborate with technology firms to create smart infrastructure.
  - **ROI Modeling**: Businesses should develop detailed ROI models which take into account the long-term benefits such as reduced downtime, improved quality and cost savings as a result of automation. These types of models are also supposed to consider intangible returns in terms of increased customer satisfaction and brand names which play a significant role in evaluating the actual value of the investment.

- **Advantages of Breaking the Financial Barriers**
  - **Cost Savings**: Cost savings may be achieved in the long term by overcoming financial barriers through streamlining of operations, minimization of waste, and high productivity.
  - **Funding Opportunities**: Those companies who effectively prove the ROI possibility of Industry 5.0 technologies might have access to funding opportunities, government incentives, or collaborations with technology providers to cover the initial costs.

iii. **Cultural Resistance: Management of Change and Acceptance of Work Force**

Another main issue is the cultural change that will be needed to be able to immerse fully in Industry 5.0 technologies. Both employees and organizational leaders can be opposed to the changes caused by automation, AI, and robotics, especially if they are afraid of losing their jobs or having their workflows broken.

- **Change Management**

- **Job Displacement**: Fear of job displacement has been one of the primary issues of workers because they believe that automation and AI will take over their jobs. Such fear may translate into resistance, especially when there is no actual plan on how the workforce is going to be affected and how their jobs are going to change.
- **Change in the Work Dynamics**: With the advent of collaborative human–machine workplaces offered by Industry 5.0, employees will have to work with high-tech devices. Such a transition can be perceived as a skeptical or even unwilling one in case workers are not aware of the advantages of cooperation with robots or AI.
- **Leadership and Vision**: Industry 5.0 technologies need strong leadership to deal with the process of change in the organization. The leaders have to make clear how the new technologies will enhance business results and the role played by the workforce in the process.

- **Workforce Acceptance**
  - **Upskilling and Reskilling**: Resistance to change can be eliminated by providing companies with training programs that enable workers to acquire skills that will be needed to operate new technologies. This is not only in technical skills but also in soft skills such as problem-solving and creativity.
  - **Trust**: Industry 5.0 is human-centric, which implies that technology is only supposed to be an addition to the workers but not a substitute. The companies should develop a culture of trust where the workers will realize that automation is meant to enhance their position and not to kill it.
- **Solution**
  - **Upskilling and Reskilling Programs**: Companies ought to invest in a program of comprehensive training, the central idea of which is to train employees in new positions, especially those that will require working with machines (e.g., AI management, robotics programming). The learning should be continuous, and the workforce must be able to keep up with the changes in technological needs.
  - **Transparency and Participation**: The leadership must be capable of communicating transparently on the benefits of Industry 5.0 and also demonstrate that robots like cobots are also designed to assist employees instead of replacing them. Involving the employees into the planning and execution processes, addressing their fears, and showing how their functions can be altered as well as the technology will establish trust and reduce opposition.
  - **Job Redesign**: This can be achieved by redesigning the job instead of laying off workers, the corporation can restructure it in such a way that it becomes more meaningful, creative and uses human capabilities in areas that demand emotional intelligence, problem solving and creativity. Human–machine collaboration, in which people and machines collaborate to increase productivity, should be encouraged.

- **Benefits of Overcoming Cultural Resistance**
  - **Improved Worker Morale**: When workers feel empowered and encouraged to change with the adoption of new technologies, they are likely to accept change and become innovative.
  - **Enhanced Teamwork**: An increased level of collaboration between people and machines can help companies unleash the potential of a technology of Industry 5.0 [10].

iv. **Policy and Regulation: Lapses and Uncertainty of the Laws**

The legal environment of Industry 5.0 technologies is emerging, and there is a great lack of legislation and regulations on how to use AI, robotics, data privacy, and cybersecurity. Such uncertainties may also delay the use of the new technologies since businesses fear being on the wrong side of the law.

- **Legal and Regulatory gaps**
  - **Data Privacy**: Since companies store and process enormous volumes of personal data using IoT gadgets, wearable sensors, and AI tools, the absence of transparent rules regarding data privacy and security may be a serious threat. As an example, the GDPR or any other data protection regulation is still a major issue regarding how personal data should be processed, transferred, and secured by most organizations.
  - **AI Ethics**: AI is linked with numerous benefits, however, its use raises ethical issues in the form of discrimination and responsibility, in addition to judgment. The lack of the global institutions to ensure that AI is developed and deployed simultaneously in a responsible manner still exists.
  - **Liability**: The issue on liability in the event of failures or accidents will require an answer since the robots and the AI systems will be more autonomous. The current laws do not take into account the complexities of the human–robot interactions in the majority of cases.
- **Regulatory Uncertainties**
  - **Technology-Specific Laws**: The majority of regulations regulating automation and digital technologies not only are outdated but can also fail to keep pace with the pace of change in Industry 5.0. This creates the question of conformity particularly where there is a difference in rules between nations or regions.
  - **Global Alignment**: This lacks any standard regulation in different regions globally and this presents a problem to companies that are operating across borders. As an example, the EU has highly regulated data privacy regulations such as the GDPR, and other nations might not have equivalent protection, and this makes cross-border activities difficult.
- **Solution**
  - **Cooperation with Policymakers**: The governments, industry leaders, and regulatory bodies should come together in order to establish clear policies

on AI ethics, data privacy, and cybersecurity. The current legal regulations, including General Data Protection Regulation (GDPR) in Europe, are to be expanded and adjusted to the new environment of Industry 5.0 technologies.
  - **Ethical Guidelines for AI**: The companies are required to create and follow an ethical code of application of AI, whereby the bias is reduced, and data privacy is held as a priority. This may comprise routine audits and openness in the decision-making process of AI.
  - **International Standards**: To make data sharing, AI usage, and robotic systems to be adopted cross-border, international standards have to be developed. This will help in lowering regulatory obstacles to the use of Industry 5.0 technologies among various nations, particularly among multinational companies.

- **Advantages of Policy Barrier Removal**
  - **Clear Compliance Pathways**: The apparent and overall compliance paths will be made available so that the business would easily and legally proceed with the adoption of the 5.0 industry technologies in order to avoid the threat of the penalty.
  - **Increased Innovation**: With more positive policies implemented driving innovation, businesses will have more motives to access new technologies and this will lead to the continued advancement of AI, robotics, and sustainability.

v. **Data Analytics, Cybersecurity, and Privacy Concerns**

Industry 5.0 is linked to the proliferation of data-driven decision-making and interconnected systems, which is dangerous because of cyberattacks, data leaks, and the breach of privacy. The operational and personal information that the IoT devices, wearables, and the AI systems collect is enormous, and it raises some concerns over data security and data privacy protection.

- **Solution**
  - **Strong Cybersecurity**: Company must invest in effective cybersecurity mechanisms that would protect IoT systems, data networks, and Artificial Intelligence systems against nefarious attacks. This involves encryptions, firewalls, multi-factor authentication (MFA), and a zero-trust security model to avoid unauthorized access.
  - **Privacy by Design**: In order to deal with the issue of privacy, businesses are supposed to adopt the process of privacy-by-design, where data protection is incorporated into systems at the initial stages. This involves anonymization of sensitive data, collection only on the required information, and giving the customers control over their data.
  - **Transparency and Accountability**: The companies are expected to be transparent with the consumers regarding the use of their data and make sure that they are following the rules of data protection such as the GDPR. Data access, data sharing, and data ownership policies ought to be clear.

vi. **Financial Sustainability and Return on Investment in Sustainable Practices**

Energy-efficient technologies, sustainable manufacturing, and circular economic systems are some of the examples of sustainable practices implemented in Industry 5.0, which may be viewed as costly at the beginning of the implementation, and the ROI is uncertain. Companies might be reluctant to invest in green techs unless it is evident that they do not have financial gains.

- **Solution**
  - **Long-Term ROI Models**: To demonstrate the business case behind the investment in sustainable technologies, companies ought to create long-term ROI models, which take into consideration energy savings, regulatory compliance (i.e., carbon taxes), brand reputation, and market demand for sustainable products. Short-term costs may be very high, but the long-term benefits may render such investments a very lucrative one.
  - **Government Incentive and Assistance**: Governments can support businesses through grant, tax credit and incentive to use green technologies. These incentives can help to offset the economic price of adopting sustainable practices and this is why companies will be more disposed to investing in sustainability.
  - **Green Tech Providers Partnership**: Under partnership, companies are in a position to spread the risk and reward part of sustainability practices with green technology parties. In the short term, the joint ventures will be effective in saving costs and accelerating the green manufacturing process.

vii. **The Industry 5.0 Technologies in Terms of Scalability and Flexibility**

Scaling Industry 5.0 systems, such as AI systems, robotic solutions, and IoT infrastructure, in large organizations or even across locations might not be easy. The firms might have a problem ensuring that technologies are flexible and scalable to various operational requirements.

- **Solution**
  - **Modular Technologies**: The companies ought to be able to acquire modular technologies that can be scaled up or down with ease according to the size and the complexity of the operations. This will enable businesses to begin small by initiating pilot projects and gradually scale up as they get to perfect their processes and have confidence in the technology.
  - **Cloud-Based Platforms**: The cloud-based platforms have the potential to make scaling Industry 5.0 technologies much easier since it does not require building on-premises infrastructure and allow accessing data and analytics remotely. Cloud solutions: They allow scaling up and down as needed.
  - **Integration of AI and IoT**: AI and IoT systems must be scalable by being made to work on various platforms and devices, and thus, they can be easily integrated in various locations and departments. It is possible to retrieve and process the data gathered from different sources in a centralized data hub that allows consistency and flexibility in decision-making [11].

Fig. 9.3 Challenges in the adoption of Industry 5.0 technologies

Figure 9.3 features various 3D icons symbolizing obstacles to the widespread adoption of Industry 5.0 technologies. The major obstacles that are described are technological complexity, financial expenses, interoperability, and regulatory ambiguities. The warning triangle indicates the seriousness of the barriers to be tackled, the financial and policy-related issues represented by the dollar sign and by the gavel. The puzzle pieces and microchips symbolize the integration and technological component that requires to be resolved to make the innovations, such as AI, robotics, and IoT, more easily applicable to the industrial context.

The challenges and difficulties of the implementation of Industry 5.0 technologies can be addressed with the help of the multi-dimensional approach that means strategic investments and good leadership, as well as efficient communication and strong technological infrastructure. Companies can position themselves well to thrive in the increasingly automated and data-driven world through technological constraints, monetary constraints, cultural pushback, policy and regulatory loopholes, and cybersecurity risks. The easy way to transition to Industry 5.0 is to find effective solutions that will include standardization, collaborative partnerships, and continuous workforce development, which will open new opportunities to grow, innovate, and become sustainable.

The governments can facilitate the process of switching to Industry 5.0 by providing financial support to companies that use green technologies or engage in the process of sustainable production. Some of the ways in which governments can assist in making the shift toward more advanced and human-oriented technologies more financially feasible include tax breaks, research grants, and funds to upskill the workforce.

9.6 The SWOT Analysis

- **Strengths**: The industry 5.0 technologies have more efficiency, customization, and cooperation between the man and the machine. All these strengths can be used to spur productivity and innovation in any industry [12].
- **Weaknesses**: The initial costs are high, integration may be difficult, and ROI is uncertain, which may hinder the adoption. Also, cultural resistance to change and retraining of the workforce are some of the factors that act as obstacles to a smooth transition.
- **Opportunities**: Among other areas, industry 5.0 will offer many opportunities for personalized products, sustainability, and empowerment of the workforce. It also creates new markets for AI-based solutions, intelligent production, and green technologies.
- **Threats**: Uncertainties in regulation, threats of cybersecurity, and legal issues regarding privacy and liability may hurt the growth of Industry 5.0 technologies. Also, there are chances that the loss of jobs through automation will result in a backlash.

Figure 9.4 presents a modern, three-dimensional interpretation of a SWOT analysis. At the center of the image is a white cube labeled "SWOT," with each surrounding quadrant in distinct colors: light blue for Strengths, red orange for Weaknesses, green for Opportunities, and orange for Threats. Each quadrant is accompanied by an icon that visually represents the concept it denotes. The diagram offers a clear, visually appealing presentation of the SWOT framework [13, 14].

Fig. 9.4 3D depiction of SWOT analysis

9.7 Learning Outcomes

By the end of this chapter, the readers will be in a position to:

- Find out the main technological, financial, cultural, and regulatory obstacles to the implementation of Industry 5.0 technologies.
- Be familiar with how complex integration is when moving on to advanced technologies after having used legacy systems and how interoperability can be resolved.
- Evaluate the economic issues of adopting Industry 5.0, such as high initial investment, long-term investment, and ROI risks.
- Learn the value of change management in managing workforce resistance, upskilling, and reskilling of workers.
- How are regulatory frameworks affecting the adoption of Industry 5.0, and why should there be global alignment in data privacy laws, AI ethics, and laws on liability?
- Devise ways of eliminating the obstacles to the adoption of Industry 5.0 by using SWOT analysis and change management.

9.8 Review Questions

1. (a) What are the major technological challenges to the implementation of Industry 5.0 technologies?

   (b) What can be done by organizations to overcome the challenges associated with legacy systems and integration of systems?
2. What is the effect of financial constraints on the adoption of Industry 5.0, especially on small and medium-sized enterprises (SMEs)?
3. (a) The workforce tends to resist industry 5.0 technologies on the grounds of culture. Which are the main sources of resistance?

   (b) What change management solutions can be adopted by an organization to make the workforce accept and cooperate with new technologies?
4. (a) Data privacy, AI ethics, and liability regulatory uncertainties are major issues when it comes to the adoption of Industry 5.0. How can organizations go through these uncertainties?

   (b) How can they ensure that they meet the changing regulations?
5. (a) Explain how government supports the collaboration of the public and the private sector to address financial and technological challenges to the implementation of Industry 5.0.

   (b) What are the ways these partnerships can speed up the use of new technologies?
6. (a) What is meant by data silos in Industry 5.0?

(b) What are the challenges posed by data silos to the realization of interoperability?

(c) What are the solutions that can be put in place to ensure there is a smooth flow of data in the systems?

7. (a) How might Industry 5.0 help to make manufacturing processes more efficient and customized?

   (b) What are the impediments to its more universal adoption in the manufacturing industry, including its high cost of initial installation and integration into the existing system?
8. What is the way by which organizations can utilize SWOT analysis to evaluate their preparedness against the adoption of Industry 5.0?
9. What are the most important regulatory issues that organizations have to address in implementing Industry 5.0 technologies, particularly on AI and data privacy? Explain the role that regulatory advocacy can take in the future of Industry 5.0.
10. (a) Write about the possible job replacement issues that are related to Industry 5.0 automation.

    (b) What can firms do to make sure that employees do not get stranded by undertaking upskilling and reskilling programs to ease transitions?

Appendix: Solution Manual

1. **(a) What are the major technological challenges to the implementation of Industry 5.0 technologies?**

Solution: The key technological barriers include the complexity of integrating new systems with legacy infrastructure and achieving interoperability between disparate platforms.

(b) What can be done by organizations to overcome the challenges associated with legacy systems and integration of systems?

Solution: Organizations can overcome these challenges by investing in system upgrades, adopting open standards for communication, and implementing middleware to bridge older systems with new technologies. Industries like healthcare and manufacturing often face difficulties due to outdated equipment and software.

2. **What is the effect of financial constraints on the adoption of Industry 5.0, especially on small and medium-sized enterprises (SMEs)?**

Solution: SMEs often face financial barriers due to the high initial costs of adopting Industry 5.0 technologies, including robotics, AI, and IoT. Companies must consider long-term ROI by evaluating benefits like improved productivity, quality control, and cost savings in maintenance. For example, predictive maintenance powered by

AI may reduce downtime and extend equipment lifespan, offering a solid ROI over time.

3. **(a) The workforce tends to resist Industry 5.0 technologies on the grounds of culture. Which are the main sources of resistance?**

Solution: Cultural resistance often stems from fears of job displacement and the disruption of established workflows. Change management strategies should focus on clear communication from leadership, transparency regarding workforce impacts, and providing upskilling and reskilling opportunities.

(b) What change management solutions can be adopted by an organization to make the workforce accept and cooperate with new technologies?

Solution: Companies can involve employees in the design and implementation of new technologies to foster trust and collaboration.

4. **(a) Data privacy, AI ethics, and liability regulatory uncertainties are major issues when it comes to the adoption of Industry 5.0. How can organizations go through these uncertainties?**

Solution: Regulatory uncertainties in data privacy and AI ethics can create obstacles for businesses looking to adopt Industry 5.0 technologies. Organizations can navigate these challenges by ensuring compliance with existing regulations like GDPR, advocating for clearer AI ethics guidelines, and staying informed about upcoming changes in the regulatory landscape.

(b) How can they ensure that they meet the changing regulations?

Solution: Organizations can navigate these challenges by ensuring compliance with existing regulations like GDPR, advocating for clearer AI ethics guidelines, and staying informed about upcoming changes in the regulatory landscape. Collaborating with policymakers can help shape future regulatory frameworks.

5. **(a) Explain how government supports the collaboration of the public and the private sector to address financial and technological challenges to the implementation of Industry 5.0.**

Solution: Government support, including subsidies, tax incentives, and infrastructure development, can reduce the financial burden of adopting Industry 5.0.

(b) What are the ways these partnerships can speed up the use of new technologies?

Solution: Public–private partnerships can provide funding for R&D and create collaborative platforms that accelerate the development and deployment of new technologies. These collaborations are crucial for reducing the cost barriers that prevent widespread adoption.

6. **(a) What is Meant by Data Silos in Industry 5.0?**

Solution: Data silos occur when data is isolated within specific systems or departments, preventing a holistic view of operations.

(b) What are the challenges posed by data silos to the realization of interoperability?

Solution: To solve this, organizations can implement enterprise-wide data integration strategies using data lakes, cloud computing, and open-source platforms.

(c) What are the solutions that can be put in place to ensure there is a smooth flow of data in the systems?

Solution: Establishing standardized protocols for data exchange ensures seamless interoperability and improved decision-making.

7. **(a) How might Industry 5.0 help to make manufacturing processes more efficient and customized?**

Solution: Industry 5.0 enables more efficient manufacturing through advanced technologies like AI, cobots, and digital twins. However, high upfront costs and integration challenges can limit adoption.

(b) What are the impediments to its more universal adoption in the manufacturing industry, including its high cost of initial installation and integration into the existing system?

Solution: High upfront costs and integration challenges can limit adoption. To overcome this, manufacturers must prioritize investments that offer long-term benefits, such as reduced downtime, enhanced quality control, and improved worker safety. A phased approach to implementation can help ease the transition.

8. **What is the way by which organizations can utilize SWOT analysis to evaluate their preparedness against the adoption of Industry 5.0?**

Solution: SWOT analysis can help companies assess their readiness for Industry 5.0 adoption by identifying strengths (e.g., advanced technical capabilities), weaknesses (e.g., outdated infrastructure), opportunities (e.g., market demand for customization), and threats (e.g., regulatory challenges). For instance, a company might use SWOT to recognize that their strong R&D capabilities are a strength, but the lack of skilled workers for AI implementation is a weakness, and they can use this insight to plan for workforce training.

9. **What are the most important regulatory issues that organizations have to address in implementing Industry 5.0 technologies, particularly on AI and data privacy? Explain the role that regulatory advocacy can take in the future of Industry 5.0.**

Solution: Regulatory challenges include compliance with data privacy laws (such as GDPR) and ensuring the ethical deployment of AI. Regulatory advocacy involves engaging with policymakers to create clear, industry-specific guidelines for AI and data privacy. By participating in discussions about AI ethics and data protection, organizations can help shape a regulatory environment conducive to Industry 5.0 adoption.

10. **(a) Write about the possible job replacement issues that are related to Industry 5.0 automation.**

Solution: Job displacement concerns can be addressed by investing in upskilling and reskilling programs that help workers transition into new roles created by automation and AI.

(b) What can firms do to make sure that employees do not get stranded by undertaking upskilling and reskilling programs to ease transitions?

Solution: Companies should focus on educating employees about the benefits of automation, such as reduced manual labor and enhanced creativity, while providing training in areas like AI management, robotics, and problem-solving to ensure workers' long-term employability.

Bibliography

1. Adel A, Alani NHS (2024) Human-centric collaboration and Industry 5.0 framework in smart cities and communities: fostering sustainable development goals 3, 4, 9, and 11 in society 5.0. Smart Cit 7(4):1723–1775
2. Zulkifli A, Wasesa M (2024) Industry 5.0 research in the sustainable information systems sector: a scoping review analysis. Int J Adv Data Inform Syst 5(2):136–156
3. Khan WA et al (2025) Industry 5.0 computer aided engineering design and manufacturing: a fourth industrial revolution perspective. Springer, Cham, pp 523–535
4. Hammad M et al (2025) Future trends and innovations in smart manufacturing. Smart manufacturing blueprint: navigating Industry 4.0 across diverse sectors. Springer, Cham, pp 413–438
5. Shabur MA, Shahriar A, Ara MA (2025) From automation to collaboration: exploring the impact of industry 5.0 on sustainable manufacturing. Disc Sustain 6(1):341
6. Kaswan MS, Rathi R, Garza-Reyes JA (eds) (2025) Industry 5.0: concepts and strategies for digital transformation. CRC Press, New York
7. Khan WA et al (2025) Pillars of Industry 4.0 and CAD/CAM. Computer aided engineering design and manufacturing: a fourth industrial revolution perspective. Springer, Cham, pp 509–522
8. Mukherjee AA, Raj A, Aggarwal S (2023) Identification of barriers and their mitigation strategies for industry 5.0 implementation in emerging economies. Int J Product Econ 257:108770
9. Tanjung T et al (2025) Drivers and barriers to industrial revolution 5.0 readiness: a comprehensive review of key factors. Green Technol Sustain 3(4):100217
10. Laddha S, Agrawal A (2024) Unveiling barriers to Industry 5.0 adoption in supply chains: a DEMATEL approach. RAUSP Manag J 59(2):123–137
11. Santos B, Costa BLC, Santos L (2024) Cybersecurity in Industry 5.0: open challenges and future directions. In: Proceedings of the 2024 21st annual international conference on privacy, security and trust (PST). IEEE
12. Basulo-Ribeiro J, Teixeira L (2024) The future of healthcare with industry 5.0: preliminary interview-based qualitative analysis. Fut Internet 16(3):68

13. Hammad M et al (2023) Framework for the implementation of smart manufacturing systems: a case in point. Processes 11(5):1436
14. Ordieres-Meré J, Gutierrez M, Villalba-Díez J (2023) Toward the industry 5.0 paradigm: increasing value creation through the robust integration of humans and machines. Comput Ind 150:103947

Chapter 10
Case Studies and Real-World Applications

10.1 Objectives

- To find practical examples of Industry 5.0 in different spheres, such as manufacturing, healthcare, agriculture, and smart cities.
- To explore the best case-studies of world leaders like Siemens, Tesla, BMW, and Amazon and to analyze how they incorporated the use of advanced technologies to improve efficiency, sustainability, and customization.
- To find out the main success factors and the pitfalls in the implementation of the industry 5.0 technologies.
- To measure the industry-specific issues and approaches to breaking the barriers to the mainstream adoption of Industry 5.0.
- To offer guidelines to businesses to implement the industry 5.0 principles by benchmarking and the best practice model.

10.2 Introduction

The shift to Industry 5.0 is just getting started. However, even a handful of early-adopted companies and industries are starting to show the promise and practical effects of the new industrial age. The analysis of these examples can provide businesses with a useful understanding of the way to implement and implement industry 5.0, solutions prioritizing the human–machine partnership, sustainability, and innovation. In this section, we will discuss some of the most successful implementations of Industry 5.0, what was successful with them, and what were the challenges encountered [1].

M. Hammad et al., *Industry 5.0: The Human-Centric Future of Industry*, Springer Series in Advanced Manufacturing, https://doi.org/10.1007/978-3-032-12622-1_10

10.3 Global Leaders: Portraits of Innovative Business and Projects

Some of the world leaders in different industries are at the forefront of the adoption of Industry 5.0. These firms are welcoming the digitalization of the latest technologies and the addition of human ingenuity, placing the emphasis on sustainability, individualization, and human–machine interactions. The strategies and innovations of these organizations can teach other organizations the most effective method of adopting AI, robotics, IoT and digital twins in their operations to become more efficient and innovative and guarantee sustainable growth.

10.3.1 Siemens: The Innovation Leader of Smart Manufacturing and AI Integration

- **Overview**: Siemens is a global leader in industrial automation and digitalization and a leader in the implementation of ideas about Industry 5.0. The company strategy is constructed on the basis of the mixture of human experience and modern automation which allows optimizing the work, increasing the level of production and promoting innovations. Siemens has been in the frontline to develop intelligent manufacturing systems that combine AI, robotics, and cobots to complement but not replace human workers.
- **Innovation**: One of the brightest examples of how Siemens is pursuing the idea of the industry 5.0 is its Amberg Electronics Plant in Germany. The plant is a very sophisticated plant where man and machines work hand in hand in a very smooth collaboration. The factory has collaborative robots (cobots) which do repetitive jobs like assembly, and human workers do higher-level jobs like problem-solving, creative design and quality control. AI-enabled production lines and digital twins will drive this human–machine partnership as they build a virtual version of the production process to optimize it in real time. The introduction of cobots will lead to a more engaged workforce and higher quality of the products produced as employees will be able to concentrate on the tasks that take creative decision-making.

  Key Features

  - Cobots that collaborate with people to perform manual and repetitive duties.
  - Digital twins and AI can be used to optimize production in real-time and predict accidents.
  - Improved human judgement and problem-solving ability in high-value activities.

- **Sustainability**: Siemens is also at the forefront in regard to sustainable manufacturing. The focus on energy-efficient production systems and smart grids has also made the company cut down on energy consumption in its operations by a great margin. Siemens has incorporated renewable energy resources and smart energy management systems into its premises, and this has assisted it in aligning its business processes with the sustainability objectives that are core to Industry 5.0. With the help of AI and IoT systems to keep track of energy consumption, Siemens can make the most out of energy consumption and minimize waste, achieving its environmental goals [2].

Sustainability Goals

- Production systems that are energy efficient and minimize the use of carbon.
- Installation of smart grids that would maximize energy consumption in Siemens facilities.
- Devotion to waste reduction and the principles of a circular economy in the operation.

- **Lessons Learned**: The experience of Siemens in implementing the 5.0 industry practices offer good lessons to other companies:
  - The cooperation of humans and machines is crucial to productivity as well as innovation.
  - The combination of AI-powered solutions and cobots allows the development of a dynamic and versatile manufacturing environment.
  - Sustainability may be properly imposed into the production system through smart tools of energy management and renewable sources of energy.
  - The data-based decision-making process allows making optimizations in real-time and giving predictive information, which is much more efficient and sustainable [3].

10.3.2 Volvo: Enhancing the Human and Machine Interaction in the Car Industry

- **Overview**: Volvo Cars is a pioneer in the use of robotics and AI along with human labor in the automobile industry. Human–machine collaboration is one of the key aspects of the company, as the company is innovative in terms of its application of collaborative robots (cobots) and AI-based solutions to streamline the production process and remain close to the theme of human creativity and well-being.
- **Innovation**: The Gothenburg factory of Volvo in Sweden is one of the brightest examples of how the principles of Industry 5.0 can be efficiently applied to the automotive industry. The business uses cobots together with the employees to aid in heavy lifting or precise assembly and other strenuous work. Such collaboration improves the safety and efficiency of workers and enables humans to concentrate on more complex parts of vehicle design and quality control. The AI is also installed within the production lines to enhance efficient work by forecasting the prospective production bottlenecks and prescribing the changes in real time.

Key Features

- Employment of cobots to help in activities that involve precision and heavy lifting.
- Production lines that are based on AI and can detect and resolve possible inefficiencies.
- Prioritizing the safety of workers and job satisfaction by offering them creative positions in design and decision-making.

• **Sustainability**: Volvo has also incorporated the concept of sustainability into its production process. The company has also adopted green technologies to minimize energy consumption and carbon emissions in its operations, and it is determined to achieve carbon neutrality by 2040. Furthermore, AI and robotics are employed to maximize material handling and recycling operations and make production as resource-saving as it can be.

Sustainability Goals

- Becoming carbon neutral in 2040 by using green technologies and sustainable means of production.
- AI and robotics to reduce waste of materials and maximize energy consumption.
- Encouraging recycling and remanufacturing during the production of vehicles.

• **Lessons Learned**: The human–machine collaboration strategy of Volvo can be learned from the following:

- Cobots and AI have the potential to reduce the physical stress on employees and increase their competency to carry out creative and strategic work.
- Sustainability may be incorporated into production systems, and it allows optimizing resources and minimizing environmental impact.
- The human-centered approach to manufacturing will help guarantee that workers will be the most important part of the process despite the growing role of automation [4].

10.3.3 Adidas: Customized Manufacturing with Human–Machine Synergy

• **Overview**: Adidas is a global sportswear organization which has embraced the concept of Industry 5.0 by incorporating the high-tech production techniques along with human innovation to make customized shoes. The use of robotics and AI in the manufacturing process of the company has helped it to be able to manufacture custom products at a high level of operational efficiency.
• **Innovation**: One of the examples of AI, robotics, and human interaction integration is the Speed factory of Adidas in Germany that allows producing custom products. Robots in the factory undertake some of the tasks like cutting and sewing

but the human workforce designs and produces products in line with the tastes of the individual customers. The information regarding the customers is used by the AI-based system to design on the basis of their demands, such as comfort, performance and style.

Key Features

- Robots are used in jobs that need accuracy and uniformity.
- Final assembly and design customization are the tasks of human workers.
- Personalization through AI makes it possible to produce custom footwear on demand.

- **Sustainability**: Adidas is also pro-sustainability as it incorporates recycled materials in its products, and it has an energy-efficient production process. The Speed factory will have a lower carbon footprint and wastage created in the manufacturing process, as the AI-powered optimization allows for reducing inefficiencies.

Sustainability Goals

- Recycled materials used in the manufacturing of footwear and apparel.
- The dedication to the minimization of environmental impact by producing energy-efficiently.
- The on-demand manufacturing model minimizes overproduction and waste.

- **Lessons Learned**: The strategy of Adidas demonstrates that Industry 5.0 can encourage innovation as well as sustainability:

- Human–machine cooperation provides the possibility of customization without compromising the efficiency of operations.
- Robotics and AI allow companies to manufacture customized goods in large quantities, which is in line with customer preference.
- The business model can be made to be sustainable, minimizing waste and environmental footprints.

10.3.4 Tesla: Breaking New Grounds in the Production of Electric Vehicles

- **Overview**: Tesla has become one of the global leaders in the production of electric vehicles by means of producing high-performance electric vehicles (EVs) in large volumes through the application of AI, automation, and machine learning. Tesla strategy is founded on the idea of the high level of robotics, the integration of data in real-time, and the focus on the customization and sustainability. The most prominent sign of Industry 5.0 in action is its Gigafactories, as it proves to be an efficient collaboration of human knowledge and automation in order to create very efficient and scalable production facilities.

- **Innovation**: Tesla has redefined how the car manufacturing process is done by integrating the application of AI-based systems, predictive maintenance, and robotics in its production line. To boost productivity and creativity, Tesla has managed to have a seamless integration of humans and robots in its Fremont and Shanghai Gigafactories.
- **Production**: Robotics and AI Tesla uses robots to automate production of complex vehicle parts and AI and machine learning to track the process and forecast potential failure to reduce downtimes and increase efficiency of the operations. The robotic arms are employed in carrying out tasks such as welding, painting and assembling and human beings have been deployed to carry out tasks that require critical thinking and creative decision making.

  Mass Customization and Efficiency: Mass customization of the vehicle is one of the greatest innovations of Tesla. With the help of AI-driven algorithms, Tesla can customize cars to meet the needs of customers and still be efficient and of high quality. Customizations can be in color, trim and optional features, with the production line of Tesla being of a high scale without compromise in speed or quality.

  Predictive Maintenance AI: Tesla's predictive maintenance is also based on machine learning, and it enables the company to recognize and solve possible manufacturing equipment problems before they can cause delays or malfunctions. This will guarantee that the production line is smooth with minimal breakdowns.

  Human–Machine Collaboration: The factories of Tesla use the collaboration of robots and people, with the first handling routine, high-precision tasks and the latter engaged in more strategic or creative work. Such collaboration enables increased productivity as well as the utilization of human skills and expertise towards their optimum application in fields such as problem-solving and innovation.

  Example: At the Fremont factory, people are required to monitor the activities of robots, change the parameters in real time, and make decisions that affect the general design and work of the vehicle. This brings flexibility and flexibility in production.
- **Sustainability**: Tesla is very serious about sustainability, and this fits the objectives of Industry 5.0 perfectly. The company incorporates sustainability in all its operations, including product design, manufacturing and energy consumption. The company has been keen on clean energy and environmental-friendly practices, which can be seen in the products it offers as well as its production.

  Electric Vehicles and Carbon Emissions: The electric vehicles (EVs) produced by Tesla are aimed at reducing carbon emissions by a large margin over the conventional internal combustion engine (ICE) vehicles. Tesla is helping the transportation sector go carbon-free by encouraging people to switch to EVs.

  Green Manufacture: Tesla is also environmentally friendly in its production where the company runs Gigafactories using solar energy. Solar panels and battery storage will restrict the consumption of fossil fuels, and this will make the manufacturing endeavors of Tesla more efficient in terms of power and environmentally friendly. One of the brightest examples of Industry 5.0 is Tesla Gigafactories which are all about green technologies and contribute to the sustainability of the supply chain.

Circular Economy and Recycling: Tesla is also on the hunt of recycling programs to reduce the wastes dumping and improve the product lifecycle. The company is also making investments in technologies that would enable it to recycle the materials it is using in its electric vehicles such as batteries in order to ensure that precious resources are not lost.

Examples: At Tesla Gigafactories, the company employs sustainable materials when manufacturing vehicles, and it concentrates on minimizing waste during the manufacturing process. This method helps to promote the model of the circular economy, the aim of which is to reduce the consumption of virgin resources and increase the process of recycling and reuse.

- **Lessons Learned**

The positive experience of Tesla in applying the principles of Industry 5.0 has a few lessons to offer other companies:

- **Human–Machine Collaboration**: AI, robotics, and human employees working at Tesla offer an example of how a cobot and automation can be applied to support human creativity and the decision-making process, not to substitute it. Collaboration between humans and robots makes work more efficient and innovative.
- **Predictive Maintenance**: The implementation of AI-based predictive maintenance makes sure that Tesla cannot take its Gigafactories offline. This shows the significance of using data to make decisions in streamlining the manufacturing processes.
- **Sustainability Integration**: Tesla has shown its commitment to sustainability through electric vehicles through solar-powered factories, and this is how sustainability may be integrated into all levels of the production chain. Industry 5.0 promotes the use of green technologies by businesses, which minimizes energy use, carbon footprint, and waste.
- **Artificial intelligence and robotics**: Tesla also demonstrates the potential of AI and robotics to support mass customization, which is why the company has been able to offer customization and efficiency at the same time. It is one of the main advantages of Industry 5.0 as AI-based solutions enable having a more flexible product without compromising the efficiency of operations.

10.3.5 BMW Group

Robotics and Human Ingenuity BMW Group is a company that has successfully integrated the use of robotics and creativity in human beings.

- **Overview**: BMW has also led in the implementation of human-oriented production systems in their factory 56 in Germany, a clear indication of how human and machine interaction can increase efficiency and customization in the production processes. The factory has collaborative robots (cobots) that collaborate with human workers to automate the production process without losing track of innovation, personalization, and sustainability. The strategy of BMW integrates the most advanced robotics with human imagination to facilitate an atmosphere of production in which both people and robots excel in their tasks.
- **Innovation**: The Factory 56 of BMW is an early stage of Industry 5.0 ideals, in which a flexible production line is capable of rapidly adapting to other car models and variants. This enables BMW to manufacture personalized cars on a large scale without compromising efficiency. Cobots are used to repeat monotonous activities like welding and assembly, whereas human employees do those things which need creativity, problem-solving, and decision-making. This flexibility gives the company the opportunity to satisfy the increasing consumer need for customized vehicles, yet it is very efficient.

  Flexible Production Lines: The production lines of BMW are flexible. This versatility allows the company to move between the various models without any stoppage, hence mass customization of vehicles that satisfy the individual needs of customers. The application of cobots helps increase the quality of assembly and precision and leaves the human to concentrate on the creative part of design and production.

  Human–Machine Synergy: Cobots collaborate with people and help them with activities that are precision and repetition-based, like fitting parts or welding components. Through this cooperation, man can undertake jobs that involve judgment, creativity, and strategy making, hence being able to control the production process, control product quality, and make decisions to make alterations in the design.

  Example: In Factory 56, cobots work on welding processes, which are usually monotonous and need accuracy. Meanwhile, human workers are engaged in work connected with creativity, e.g. adjustment in design or elimination of problems appearing on the production line.
- **Sustainability**: BMW has also employed sustainability in its production through the use of digital twins. Simulation of various production scenarios and optimization of energy consumption and resource use in production is possible with these digital replicas of physical systems provided by BMW. Through virtual modelling and testing of production processes, BMW will be able to minimize wastage and poor efficiency before it hits the factory floor.

Carbon-Neutral Manufacturing: BMW strives to be carbon-neutral in its manufacturing process by the year 2030, and Factory 56 will be at the center of the process. The renewable energy and energy-efficient systems used in the factory minimize the carbon footprint of the factory and also complement the wider sustainability of BMW as a company.

Energy Optimization: Energy optimization can be achieved by using digital twins, and BMW can monitor energy consumption in real-time, enabling it to optimize its energy consumption and minimize wastage within its production systems. This is in keeping with another bigger objective of the company to reduce its impact on the environment and towards a more circular economy.

Sustainable Materials: BMW is also concerned with sustainability in terms of using recycled materials and incorporating elements that are environmentally friendly in their cars. This will make sustainability to be integrated into the whole process of manufacturing and product life cycle.

10.3.6 Revolutionizing Logistics and Supply Chain Operations

- **Overview**: Amazon, being one of the largest leading companies in e-commerce and supply chain management, has disrupted logistics with the help of AI-powered warehouses, robotic solutions, and real-time data analytics. The Amazon model combines AI, robotics, IoT, and human labor to establish a very efficient and smooth fulfilment mechanism that is capable of delivering its massive operations worldwide. The firm has shown how highly sophisticated automation can be used together with human supervision to establish a swift, precise and effective logistics system.
- **Innovation**: Amazon is applying robots so as to handle the inventory, pick the items and make orders both by human workers. The demand can be predicted and the levels of the stock can be optimized by using AI algorithms, whereas the entire supply chain will be monitored in real time with the help of IoT devices that will provide data that is significant in making the decision and effectiveness of the operations.

  Artificial Intelligence Forecasting: Amazon exploits the strength of Artificial Intelligence to come up with forecasts and inventory management. These algorithms help in ensuring that the products are stocked and are available to be picked when needed, thereby increasing the accuracy of the orders and the time of delivery. This kind of predictive capability helps Amazon to be very responsive to changes in consumer demand, and its supply chain is highly dynamic and strong.

  Warehouses: The Amazon fulfilment centers feature the use of robots to assist human workers by carrying products, picking items, and helping to sort. Such robots are used to work with human workers who are required to focus on the tasks that imply decision-making, quality control, and finishing. This combined supply channel increases efficiency, reduces the labor cost and increases accuracy in meeting orders.

Example: Amazon has Kiva robots that work in its fulfilment centers and carry products and deliver them to picking stations. The human employees are concerned with sorting and packing the items so that the final product would arrive to the consumer fast and with the right accuracy.

- **Sustainability**: Amazon is determined to reach net-zero carbon emissions by 2040 and has already begun to apply sustainable logistic solutions. Sustainability in the company is aimed at minimizing the carbon footprint, renewable energy, and transportation optimization.

  Electric Delivery Vehicles: In a bid to be green, Amazon is incorporating electric delivery vehicles in its fleet. This minimizes the emissions caused by last-mile deliveries and assists the company in becoming carbon-neutral.

  Automated Sorting Systems: Amazon has installed automated sorting systems which enhance efficiency in the warehouses and minimize environmental effects due to over-packaging and transportation. Such automated systems assist in optimizing the management of the inventory and making sure that the products are delivered in the most efficient and environmentally friendly ways.

  Sustainable Packaging: Amazon has worked to minimize packaging waste through recyclable materials and the size of the packaging in every order. This is one of the initiatives that Amazon has in sustainable packaging solutions, which minimize waste and energy usage [5].

10.4 Industry-Specific Cases: Detailed Look at Successful Industry 5.0 Implementations

It is more and more implemented in various industries as the concept of Industry 5.0 can provide the power of high technologies and mix it with human imagination, collaboration, and moral responsibility. Here, one can find a list of examples of the successful implementation of the industry 5.0 principles in the sector-specific context that demonstrates the use of these technologies to achieve higher productivity, sustainability, and overall outcomes.

10.4.1 Healthcare: Robotic Surgery and Personalized Medicine

Robotic surgery systems, AI, and personalized medicine are transforming the healthcare sector in a considerable way. These innovations are assisting in better outcomes for the patients, surgical accuracy, and customized treatment that improves overall healthcare delivery.

- **Robotic Surgery**: One of the most promising representatives of robotic surgery is Da Vinci System by Intuitive Surgical. Surgeons are able to operate in minimum invasiveness and high accuracy with the aid of this system. da Vinci robot consists of robotic arms, high-definition 3D cameras, and real-time data analysis that assist surgeons to conduct their work more precisely and controlled.

  Key Features: Minimal invasive procedures result in smaller surgical incisions that translate to quicker recovery and less trauma to the patient.

  - Robotic assistance allows surgeons to do complex procedures with better accuracy.
  - The system provides real-time feedback of data, enabling surgeons to make more informed decisions during operations.

  Benefits

  - Better patient outcomes: Robotic surgery enhances accuracy and minimizes possible human error, resulting in higher recovery rates and lower complications.
  - Faster healing: Minimally invasive surgery will lead to faster recovery of the patients since the trauma caused by surgery is minimal.

  Example: Mayo Clinic has been using robotic surgery in complex surgeries related to cancer where precision is very important. Da Vinci robots help surgeons conduct an accurate operation, enhance the success of surgery, and reduce pain after surgery.
- **Personalized Medicine**: The sphere of personalized medicine is becoming more and more popular in the world of healthcare, and AI is playing a crucial role in its development. With the help of big data, genetic data, and clinical records, AI-based technologies can be used to develop highly individualized treatment plans that would suit the individual genetic makeup and medical history of the patient.

  Key Technologies

  - Patient data, such as genetic, environmental and lifestyle data, can be analyzed using AI and make better diagnoses and treatment plans.
  - Machine learning models will be able to predict the patient response to a particular intervention so that more effective interventions can be done and fewer side effects experienced.

  Benefits

- More accurate treatments: Taking into account genetic information and clinical history, one can customize treatment to the patient, which results in more successful results.
- Less side effects: Personalized medicine uses a more specific treatment for the disease, and therefore, there is less likelihood of unnecessary treatments and side effects.

Case study: Tempus is a health technology organization that applies AI to large volumes of clinical information, genetic data, and patient histories to create individualized treatment regimens for cancer patients. This allows oncologists to offer treatment depending on the genetic makeup of an individual, which enhances the success rate.

Example: Mayo Clinic is utilizing machine learning and predictive analytics to personalize cancer therapy. Mayo Clinic can propose treatment regimens according to personal specificities, including genetics and health history, by analyzing massive amounts of data about patients, which would result in more efficient and personalized treatments.

10.4.2 Manufacturing: Smart Factories and Human–Machine Collaboration

Industry 5.0 principles are successfully applied in the manufacturing sector and aim at increasing productivity, customization, and sustainability with the help of smart factories, cobots, and AI-based production systems.

- **Cobots and Smart Factories**

In Industry 5.0, smart factories make use of a blend of IoT sensors, AI, and collaborative robots (cobots) to maximize production. Being safe to collaborate with human workers, Cobots contribute to the increased efficiency of work as they can complete repetitive and physically challenging tasks, whereas human employees can concentrate on complex decision-making, design thinking, and problem-solving.

Key Features

- The AI and IoT combination enable the ability to track production processes in real time.
- Cobots are used in assembly, packaging and quality control, and humans are used in high-skill tasks like design, troubleshooting and decision-making.
- Automation has been balanced with human creativity, thus making it flexible and efficient for product customization.

Benefits

- Higher productivity: Cobots and AI systems are able to work 24 h a day, which speeds up the production process and minimizes downtime.
- Customization: Mass customization of products may be offered by manufacturers with the use of smart factories without compromising efficiency or quality.
- Human empowerment: Robots do not replace workers but empower them to work on more rewarding and creative activities.

Example: Factory 56 in Germany by BMW is one of the examples of human–machine collaboration. Cobots are employed in the factory where they assist in the assembly and welding process, and the human employees are involved with quality assurance and creative design. This will help BMW to produce customized vehicles in a cost-effective manner without compromising.

- **Sustainability in the Manufacturing**

  Industry 5.0 is also about sustainability in industry. Digital twins, AI, and automation are being used by companies to optimize energy consumption, minimize waste and enhance resource management during the production process [6].

Key Technologies

- Digital twins are virtual representations of the physical manufacturing systems that are used to analyze and optimize energy consumption, material flow and waste management.
- AI algorithms monitor the amount of energy used and the materials consumed, proposing optimizations that would minimize environmental effects.

Benefits

- Resource optimization: Digital twins and AI optimize the processes in production to reduce the consumption of energy and wastage of materials.
- Less environmental impact: AI and automation assist organizations in achieving their sustainability objectives by cutting carbon emissions, energy use and waste.

Example: Gigafactories of Tesla are the best examples of sustainable manufacturing. These factories are run on renewable energy, apply AI to energy use, and include sustainable materials in producing electric cars. Tesla is seeking to minimize its carbon footprint and still be efficient in production.

10.4.3 Agriculture: Precision Farming and Sustainable Agriculture

Industry 5.0 is also bringing an improvement in precision farming, resource and sustainability in the agriculture sector.

- **Precision Farming**: Precision farming makes use of IoT sensors, AI, and drones to follow the state of the soil crops and weather conditions in real time. This information will be utilized to make informed decisions regarding irrigation, fertilization and pest control, and this will ensure that the resources are utilized effectively, and crops are grown in the best way possible.

Key Features

- The IoT sensors measure the environmental factors, including soil moisture, temperature, and humidity, and AI use such data to streamline farming activities.
- Drones can be used to cover vast fields, check crop illnesses, and spray fertilizers or pesticides in the exact location.

Benefits

- **High agricultural output**: With the best use of resources as well as the ability to respond swiftly to environmental conditions, farmers can ensure agricultural output.
- **Resource conservation**: Precision farming saves water, fertilizer and pesticide consumption and wastage and enhances sustainability.

Example: The example is John Deere who has integrated AI and IoT into farming equipment to be able to gather real-time data and analyze it. This will empower farmers to plant more, irrigate and manage pests to a maximum hence there will be more utilization of resources and higher production of crops.

Sustainability: Precision farming will allow sustainable usage of resources, thus maintaining the long-term environmental well-being and conservation of natural ecosystems. Sustainability in agriculture is concerned with minimizing environmental degradation, resource conservation and increasing crop yield. The AI-based system is important in making farming sustainable because it helps the farmers make more informed decisions based on data to reduce waste and enhance efficiency.

Key Technologies

- Artificial intelligence and IoT sensors in soil monitoring systems allow farmers to keep track of the soil conditions in real-time and make sure that they are using the right amount of water and avoid over-irrigation.
- The use of data integration of weather stations, soil sensors, and crop monitoring tools enables the farmers to change their practices according to the current environmental conditions.

Benefits

- Minimized water consumption: Through the optimization of irrigation according to the exact level of soil moisture, farmers can save a lot of water consumption, which is especially significant in areas with water deficiency.
- Healthier crops: The real-time soil and crop monitoring will ensure that problems are detected early on before they can affect the crop yields of the farmers.

- Improved sustainability: Integration of AI in the management of resources such as water and fertilizers contributes to more sustainable farming activities as well as a smaller environmental impact of agriculture.

Case study: CropX is an Israeli firm that has created an AI-assisted soil monitoring system that entails embedding IoT sensors in the soil to maximize water consumption. The system assists farmers in making better decisions regarding irrigation, and thus, water is used only when it is required and not wasted, resulting in better crop yield.

- **Vertical Farming**

Vertical farming is the new form of urban agriculture involving hydroponics, LED light, and AI technologies that allow growing crops in layers vertically. The solution is much more sustainable compared to traditional farming because it uses a lot less land and water.

Key Technologies

- AI and machine learning can optimize the growing environment of each plant by changing the environmental parameters such as light, temperature and humidity to maximize yields.
- Hydroponic systems enable the growing of plants without soil through the application of a water solution that is rich in nutrients, consequently saving the amount of water consumed in traditional soil farming.
- The LED lights offer the perfect wavelength of light that the plants need to grow with less energy consumption.

Benefits

- Space efficiency: Vertical farming enables the production of food in the city, using otherwise unutilized areas such as rooftops or warehouses.
- Less water: Hydroponics consumes less water compared to conventional agriculture, and so it is perfect in water-scarce regions.
- Less environmental impact: Vertical farms can be fueled by renewable energy sources, and there is no need to transport them, as they are built near urban areas.

Case in point: Aero farms and Bowery Farming are some of the first companies that use AI, hydroponics, and LED lights to cultivate crops in indoor vertical farms. Such systems will not only decrease land use but also conserve water and energy, which is a sustainable means of feeding the cities.

10.4.4 Smart Cities: Strengthening the Infrastructure, Energy and Citizen Participation

The development of smart cities is also being done in the light of Industry 5.0. With the help of IoT sensors, AI, and digital twins, cities have an opportunity to optimize the infrastructure, energy management, and citizen engagement to make cities more efficient, sustainable and livable.

- **Infrastructure**

A smarter method of managing the urban infrastructure is given to smart cities with the use of digital twins, IoT sensors, and AI algorithms. By simulating the operations and infrastructure of the city, the planners of the city will be able to make decisions based on data and this will assist in optimization of the traffic flow, improve the route of the public transport system and save the consumption of energy.

Key Technologies

- Digital twins offer an urban system virtual model, and city planners can simulate situations and optimize workflows such as traffic control, energy delivery, and garbage removal.
- The sensors can be installed on the infrastructure, including roads, bridges, and buildings, to detect their condition and provide the necessary maintenance in time.

Benefits

- Better efficiency: Infrastructure management in real-time data results in less traffic congestion, better public transport, and more efficient energy consumption.
- Saving costs: Digital twins can also be used to anticipate maintenance, hence eliminating costly repairs and extending the infrastructure life.

Example: Singapore is the leader in the use of smart city technologies. Digital twins and artificial intelligence enable the city to model traffic, optimize the route of public transport and have real-time information about energy consumption to make the city more efficient.

- **Energy Management**

The AI-enabled smart grids will help cities to distribute the energy more efficiently, make the energy more efficient and introduce renewable sources of energy to the grid. Cities with real-time data can control the supply and demand, reduce waste and ensure that the energy resources are utilized.

Key Technologies

- AI algorithms can be used to predict energy demand and supply, and therefore, it is possible to redistribute energy in real-time.
- IoT and smart meters measure energy consumption throughout the city and can be used as valuable data to use resources most efficiently.

Benefits

- Energy efficiency: Smart grids aid in avoiding wastage of energy and making sure that the power is allocated based on the requirement, making the system efficient.

- Combination of renewable energy: AI permits the more effective combination of solar, wind, and other renewable energy sources, making less reliance on fossil fuel.

Example: Barcelona has launched smart grids that use AI to ensure that energy is distributed efficiently and that wastage is minimized. The city incorporates renewable sources of energy such as solar and wind power to satisfy the energy requirements, which are sustainable in aid of its environmental objectives.

- **Citizen Engagement**

Smart cities are also aiming at enhancing citizen engagement through digital platforms that help residents to be involved in the government decision-making processes, utilize public services, and give feedback.

Key Technologies

- Digital tools and mobile apps enable citizens to interact with the city services, report problems and access local events, policies, and services data.
- The AI will be utilized to analyze the feedback of people and find out the trends so that the government can make data-driven decisions concerning the development of cities.

Benefits

- More transparency: There are increased chances of citizens being involved in the decision-making process, which helps to build trust and cooperation between the state and citizens.
- Better public services: When the cities gather and analyze the views of the people, they are in a better position to coordinate their policies with the needs and priorities of the people [7].

Example: Estonia has emerged as a digital government leader and provides citizens with access to government services in a digital platform. This system allows effective provision of services and promotes the involvement of the people in the decision-making process.

10.5 Analysis: The Success Factors and Pitfalls

The road to the adoption of Industry 5.0 provides great prospects for increasing productivity, sustainability, and human–machine cooperation. Nonetheless, it is also associated with some difficulties that organizations need to maneuver. Analyzing the practical case studies and the industry-specific examples, the businesses can determine the major success factors and the potential pitfalls that might influence the successful implementation of the industry 5.0 technologies.

10.5.1 Success Factors

- **Human-Centric Design**

Human-centricity is one of the main principles of Industry 5.0. Industry 5.0 focuses on human–machine collaboration, unlike the past industrial revolutions that focused on automation to substitute human labor. It is important to ensure that humans and machines complement each other to produce an active workforce and a more productive entity.

Dominant Insights

- Effective realizations mean that the technologies created should augment and not substitute for human capabilities.
- Robotics, AI systems, and cobots need to be designed to support human creativity, decision-making and emotional intelligence to allow employees to work on high-value tasks.

Example: In BMW Factory 56, collaborative robots are used together with human workers to increase productivity and give workers the opportunity to be strategic and creative in their work, which increases job satisfaction and efficiency.

- **Good Vision and Leadership**

To achieve success with Industry 5.0, Leadership must be the force that leads to the implementation of technologies by means of a clear vision, which is related to the long-term business objectives. Effective Leadership is due to the establishment of direction that would make sure that the integration of new technologies is in line with the strategic goals of the company.

Dominant Insights

- To be successful, it must have a strategic plan with specific goals, objectives, and measures of success of the technology adoption.
- Leaders should explain why they should embrace technology to succeed in business and involve employees during the transition process.

Example: Elon Musk, as the CEO of Tesla, has been instrumental in spearheading the company to pursue innovative activities on robotics and AI in the electric vehicle manufacturing process, and so far, the company remains one of the most sustainable manufacturing companies.

- **Flexibility and Scalability**

Scalability of technologies and making them adjustable to future changes is also essential in the adoption of Industry 5.0. Companies must consider systems that can expand with business and be able to integrate with the available infrastructure. The new technologies and legacy systems have to be compatible to ensure smooth operations.

Dominant Insights

- Modular and scalable systems guarantee that the business can keep growing and developing its functionality without the necessity to upgrade the current technology.
- Flexible technologies enable businesses to adapt to market dynamics, regulatory policies, and changes in technologies, thus becoming sustainable in the long term.

Example: The Amberg Electronics Plant of Siemens has cobots and AI-based production lines, which can be scaled and reconfigured to address the changing needs of the customers. The digital twins of the factory contribute to production process optimization, which is flexible and adaptable.

- **Training and Development of Workforce**

One of the success factors of the transition to Industry 5.0 is the guarantee of providing the employees with the needed skills to collaborate with the advanced technologies. Workforce upskilling enables an easy transition and makes sure that the potential of new technologies will not be wasted.

Dominant Insights

- Companies need to invest in training activities that would enable the employees to utilize the new technologies so that they can be relevant and interested in them.
- There is also constant learning and adjustment to the new technologies that change rapidly [8].

Example: Training programs of Volvo make workers proficient in robotics and AI so they can work with cobots on the production line and concentrate on tasks that involve creativity and problem-solving.

10.5.2 Common Pitfalls

- **Inability to integrate**

The integration of new systems with the existing infrastructure can be considered one of the most tremendous challenges to the adoption of Industry 5.0 technologies. The systems might not operate efficiently without adequate interoperability, which results in inefficiency and high costs.

Dominant Insights

- Interoperability is critical to the smooth operation of both the old systems and the new systems.
- Companies are advised to make use of open standards and modular systems that can be easily integrated with the current infrastructure.

Example: The reason is that a company that does not integrate its new AI-powered tools with its current software might face issues with real-time decision-making that may result in the disruption of production and the supply chain process.

- **Underestimating the Costs**

Most companies fail to realize the start-up expenses of introducing Industry 5.0 technologies. Although the technologies could result in cost savings and efficiency in the long run, the upfront cost, which can be significant in the forms of infrastructure, training and system upgrades, can be very high.

Dominant Insights

- It is important that the organizations evaluate the costs and benefits of new technologies in the long run and the short run.
- To find out whether the investment is worthwhile and make sure that resources are distributed efficiently, a cost–benefit analysis may be conducted.

Example: Amazon encountered difficulties when it first implemented the use of robotic systems in its fulfilment centers. The cost of the initial investment in the automation and robotics infrastructure was high, but in the long run, the efficiency and cost reduction were higher than the initial investment.

- **Cultural Resistance**

One of the most widespread issues organizations have to deal with when implementing new technologies is resistance on the part of the employees. The adoption process can be hampered by fear of loss of job, lack of knowledge or just plain reluctance to change.

Dominant Insights

- Businesses need to create an atmosphere of trust and cooperation when employees can view new technologies as a means of helping them as opposed to making them redundant.
- Resistance can be overcome through clear communication concerning the advantages of the adoption of technology, training, and upskilling opportunities [9].

Example: BMW paid attention to employee engagement and training programs in order to make sure that workers did not feel that the integration of cobots meant that they would be substituted by machines. Cultural resistance was minimized by open communication and transparency, which made the adoption process easier.

10.6 Analytical Focus: Comparative Benchmarking Model and Best Practice Models

Comparative benchmarking and the use of best practice models are important to businesses that wish to effectively implement Industry 5.0. Through the investigation of the best representatives of the industry, which have already managed to incorporate Industry 5.0 concepts, the organizations can find the strategies and practices that will fit their specific requirements and goals. Benchmarking and the best practice models will assist businesses in changing towards human-centric and sustainable manufacturing processes, also remaining competitive and able to maximize the technological changes.

10.6.1 Comparative Benchmarking

Comparison of benchmarking involves a comparison of the performance of a company with the performance of other companies that have attained a leadership position in the industry. This analysis enables the organizations to assess their current position in regard to the use of technology, efficiency of their operations and sustainability. Through such identification, the businesses will find an opportunity to utilize the insights and strategies of the best-performing companies to make their businesses successful.

- **Advantages of Comparative Benchmarking**
  - Spotting Gaps: Benchmarking enables businesses to know the gaps in terms of technology adoption, efficiency or sustainability where they are likely to be behind their competitors.
  - Follow the Best Practices of the Industry Leaders: Companies can also follow the best practices of industry leaders to incorporate strategies and technologies that have already been successful.
  - Continuous Improvement: Benchmarking urges companies to keep reviewing and upgrading their own operations in a bid to remain competitive in a fast-changing market.
- **Key Insights**

  The following are the Key Focus Areas in Benchmarking:
  - **Technology Adoption**: Analyze the level of uptake of Industry 5.0 technologies such as AI, cobots, IoT and digital twins by companies. Technology adoption benchmarking can be used to determine whether the organization is on track in terms of innovation or whether it should invest in some respects.
  - **Operational Efficiency**: Evaluate efficiency by utilizing measures of performance such as downtime, speed of production, utilization of resources as well as cost-cutting to guarantee maximum productivity.

- **Sustainability**: Examine the efforts made by businesses to incorporate sustainable processes, including green technologies, recycling, energy-efficient production, and circular economy models.

Example: Tesla Gigafactories are used as an example of electric vehicle production with the introduction of AI, robotics, and energy-saving production systems. Tesla has been keen on green technologies through sustainable production and renewable energy, which makes it a good example of the implementation of green technologies.

Benchmarking Process: The industry reports case studies, and public data enable companies to compare their performance with the leading players. Periodic benchmarking activities will enable organizations to determine where they can make some improvements, especially in line with Industry 5.0 concepts such as human–machine collaboration and sustainability [10].

10.6.2 Best Practice Models

Best practice models allow learning about the strategies that have been used effectively to implement Industry 5.0 in the most prominent companies. Examining the way companies like Siemens, Tesla, and BMW have incorporated the use of advanced technologies and human-focused strategies, it is possible to adjust these practices to the needs of a particular company or organization.

- **Important Issues of Best Practice Models**

Human-Centric Collaboration: The most successful firms are focusing on human–machine cooperation rather than automation of workplaces with the replacement of human employees. Collaborative robots (cobots) can work with employees, boosting creativity, problem-solving and innovation.

Example: BMW Factory 56 has introduced cobots into the production line to help with repetitive tasks and enable human workers to perform creative and decision-making tasks to increase flexibility and customization in production.

Implementation Plan: Phase-Wise Implementation: One of the best practice models may be the implementation of Industry 5.0 technologies gradually with the initial small pilot projects and then expanding. This gradual process minimizes risk and enables businesses to evaluate the effect of the technologies before undertaking bigger investments.

Case in Point: The Amberg Electronics Plant of Siemens implemented digital twins and AI-based manufacturing lines step by step. They began by employing them in a particular part of the production line, which helped them to refine the system before implementing them on the whole scale of the plant.

Sustainability Integration: The key to the success of companies is to incorporate sustainability in their production cycle so that the implementation of Industry 5.0

would not contradict environmental and social objectives. This involves efficiency in the use of resources, minimization of waste and use of green technologies.

Example: Tesla has incorporated sustainable production processes in its Gigafactories using solar energy, energy-efficient systems, and recycling materials it uses to manufacture vehicles in a bid to reduce environmental impact.

Training and Upskilling of Workforce: The focus on workforce development as the best practice model involves training and constant learning, which would allow the employees to effectively work with AI, robotics, and other advanced technologies. This assists organizations in preventing the problem of cultural resistance as well as employee displacement [11].

Example: The training programs offered by Volvo guarantee that the employees will have the skills that are needed to work with AI-driven systems and cobots on the production line, and the process of transitioning to Industry 5.0 will be relatively easy.

- **Best Practice Implementation Strategy**

Pilot Programs: The most effective way to start is small-scale pilot projects, which will be aimed at testing the effectiveness and ROI of a particular technology, e.g. AI, robotics, or digital twins, and then scale up.

Employee Engagement: Encourage a culture of teamwork where employees participate in the implementation of new technologies. This makes it more engaging, and human talents and creativity would be at the center of the process.

Constant Assessment: Conduct periodic performance reviews and technology checks so that the industry 5.0 efforts are having the intended outcomes in regard to efficiency and sustainability as well as employee wellness.

10.7 Learning Outcomes

After going through this chapter, the reader will be able to:

- Learn how Industry 5.0 is affecting the major industries, which include manufacturing, healthcare, agriculture, and smart cities.
- Evaluate the approaches that have been used by leaders around the globe in embracing Industry 5.0 technologies and apply the same to their respective organizations.
- Determine the general issues businesses encounter when trying to combine advanced technologies with legacy systems and suggest the ways through which the problems can be addressed.
- Understand the value of a human-focused design and cooperation in the effectiveness of the implementation of Industry 5.0.

- Employ comparative benchmarking and best practice models in influencing the implementation of Industry 5.0 technologies in their organizations.
- Create a strategic plan according to which the principles of Industry 5.0 can be implemented, including both industry- and company-specific issues.

10.8 Review Questions

1. What has the manufacturing industry gained by adopting the industry 5.0 solutions of cobots, AI-powered production lines, and digital twins?
2. How is Industry 5.0 related to the healthcare industry, and how can it improve personalized medicine and robotic surgery?
3. (a) Which are the major technologies that aid precision farming in Industry 5.0?

   (b) What is the role of AI, IoT, and drones in the optimization of the use of resources and higher crop yield in agriculture?
4. What are the roles of digital twins, AI-based energy management and citizen engagement platforms in making the urban environment more sustainable and efficient?
5. (a) Which challenges are most prevalent to organizations in the implementation of Industry 5.0 technologies, in particular, technological integration, acceptance by the workforce, and regulatory compliance?

   (b) What can be done to get to these barriers?
6. (a) Examine how collaboration and customization work in Industry 5.0?

   (b) What are the roles of cobots and AI in improving the capacity of organizations to tailor products and services to the personal needs of customers without losing operational efficiency?
7. What is the role of AI, robotics and IoT technologies in smart manufacturing environments in achieving sustainability objectives?
8. (a) Discuss the role of Leadership and vision towards the successful implementation of Industry 5.0 technologies.

   (b) What can be done so that organizations can guarantee themselves a clear strategic vision and effective leaders to overcome the challenges of Industry 5.0 adoption?
9. (a) What are the success factors of implementing Industry 5.0 technologies?

   (b) Explain how scalable and flexible technologies, workforce training and good leaders help in successful implementations?
10. How can benchmarking and best practice models be used to determine the level of preparedness of companies to Industry 5.0?

Appendix: Solution Manual

1. **What has the manufacturing industry gained by adopting the Industry 5.0 solutions of Cobots, AI-powered production lines, and digital twins?**

Solution: The manufacturing sector benefits from Industry 5.0 by enabling mass customization, allowing companies to tailor products to individual customer preferences without sacrificing efficiency. Cobots and AI-driven production lines help automate repetitive tasks, while digital twins allow for real-time monitoring and optimization. Challenges include the high initial costs and the integration of these technologies with legacy systems, which may require significant investment in infrastructure.

2. **How is Industry 5.0 related to the healthcare industry, and how can it improve personalized medicine and robotic surgery?**

Solution: Industry 5.0 enhances healthcare through personalized medicine, where AI analyzes genetic data to recommend tailored treatment plans. Robotic surgery improves precision and minimizes invasiveness, leading to faster recovery times. Examples include Mayo Clinic using machine learning for cancer treatment and Intuitive Surgical's da Vinci system enhancing surgical precision. These technologies reduce costs by minimizing errors and improving patient outcomes.

3. **(a) Which are the major technologies that aid precision farming in Industry 5.0?**

Solution: Precision farming in Industry 5.0 relies on IoT sensors to monitor soil conditions and weather patterns, drones for field surveillance, and AI to optimize irrigation, fertilization, and pesticide use.

(b) What is the role of AI, IoT, and drones in the optimization of the use of resources and higher crop yield in agriculture?

Solution: These technologies help reduce resource consumption and increase crop yield, making farming more sustainable. For example, John Deere uses AI-powered tractors to automate planting and harvesting, ensuring resource efficiency and high productivity.

4. **What are the roles of digital twins, AI-based energy management and citizen engagement platforms in making the urban environment more sustainable and efficient?**

Solution: Industry 5.0 technologies contribute to smart cities by using digital twins to simulate urban planning scenarios, optimizing traffic flow, and reducing energy consumption. AI-powered energy management systems like those in Barcelona ensure efficient use of renewable energy, while citizen engagement platforms like those in Estonia foster greater public participation in government decisions. These technologies help create sustainable, livable urban environments.

5. **(a) Which challenges are most prevalent to organizations in the implementation of Industry 5.0 technologies, in particular, technological integration, acceptance by the workforce, and regulatory compliance?**

Solution: Common challenges include the complexity of integrating new technologies with legacy systems, cultural resistance to change, and the uncertainty of regulatory frameworks. Overcoming these barriers requires a clear strategy, strong leadership, and investment in workforce training.

(b) What can be done to get to these barriers?

Solution: Overcoming these barriers requires a clear strategy, strong leadership, and investment in workforce training. Interoperability can be achieved through standardization, and regulatory compliance can be ensured by engaging with policymakers and adhering to international guidelines.

6. **(a) Examine how collaboration and customization work in Industry 5.0?**

Solution: Collaboration and customization are central to Industry 5.0. Cobots and AI enable companies to create highly customized products by working alongside human workers.

(b) What are the roles of cobots and AI in improving the capacity of organizations to tailor products and services to the personal needs of customers without losing operational efficiency?

Solution: For example, in automotive manufacturing, AI and robotics work together to allow customers to personalize vehicle features while maintaining the efficiency of the production line. The challenge lies in the initial investment and the need to design flexible, scalable systems.

7. **What is the role of AI, robotics and IoT technologies in smart manufacturing environments in achieving sustainability objectives?**

Solution: The integration of AI, robotics, and IoT contributes to sustainability by optimizing resource use and reducing waste. Smart manufacturing processes reduce energy consumption and emissions. However, the challenge is balancing innovation with environmental responsibility, as new technologies may require significant energy during production and deployment. Companies must focus on energy-efficient solutions and incorporate sustainable practices into their operations.

8. **(a) Discuss the role of Leadership and vision towards the successful implementation of Industry 5.0 technologies.**

Solution: Leadership and vision are essential in implementing Industry 5.0 technologies. Leaders must communicate a clear strategy, ensure alignment between technology adoption and business goals, and manage the cultural transition within the organization.

(b) What can be done so that organizations can guarantee themselves a clear strategic vision and effective leaders to overcome the challenges of Industry 5.0 adoption?

Solution: Effective change management practices, such as involving employees in decision-making and providing training, are crucial for successful implementation.

9. **(a) What are the success factors of implementing Industry 5.0 technologies?**

Solution: Key success factors include a clear vision, scalable technologies, and workforce training. Successful companies like Siemens and Tesla prioritize flexibility in production lines and focus on upskilling employees to work alongside advanced technologies.

(b) Explain how scalable and flexible technologies, workforce training and good leaders help in successful implementations?

Solution: Successful companies like Siemens and Tesla prioritize flexibility in production lines and focus on upskilling employees to work alongside advanced technologies. These factors contribute to increased productivity, innovation, and a smooth transition to Industry 5.0 operations.

10. **How can benchmarking and best practice models be used to determine the level of preparedness of companies to Industry 5.0?**

Solution: Benchmarking allows companies to evaluate their progress against industry leaders. By comparing their technology adoption, efficiency, and sustainability efforts with companies like Siemens, Tesla, and Amazon, organizations can identify areas for improvement. Best practice models such as phased implementation and pilot projects help reduce risks and ensure a successful transition to Industry 5.0 technologies.

Bibliography

1. Rejeb A et al (2024) The research landscape of industry 5.0: a scientific mapping based on bibliometric and topic modeling techniques. Flex Serv Manufact J 14:1–48
2. Gajdzik B (2025) Digital transformation towards smart steel manufacturing. Springer, Cham
3. Khan WA et al (2025) Industry 5.0 computer aided engineering design and manufacturing: a fourth industrial revolution perspective. Springer, Cham, pp 523–535
4. Söderlund H (2025) XR enabled operator training
5. Mao Z et al (2024) Model and metaheuristic for human–robot collaboration assembly line worker assignment and balancing problem. Comput Oper Res 165:106605
6. Yang J, Liu Y, Morgan PL (2024) Human–machine interaction towards Industry 5.0: human-centric smart manufacturing. Dig Eng 2:100013
7. Adel A, Alani NHS, Jan T (2024) Factories of the future in industry 5.0—softwarization, servitization, and industrialization. Internet Things 28:101431
8. Rejeb A et al (2025) Industry 5.0 as seen through its academic literature: an investigation using co-word analysis. Discov Sustain 6(1):307
9. Bradford N, Bramwell A (eds) (2014) Governing urban economies: innovation and inclusion in Canadian city regions. University of Toronto Press, Toronto

10. Saniuk S, Grabowska S, Adamik A (2024) Smart organizations in Industry 5.0: a human-centric approach. Taylor & Francis, New York
11. Carayannis EG, Morawska J (2023) University and education 5.0 for emerging trends, policies and practices in the concept of industry 5.0 and society 5.0—Industry 5.0: creative and innovative organizations. Springer, Cham, pp 1–25

Chapter 11
Future Outlook and Trends

11.1 Objectives

- To investigate the upcoming technologies, i.e., quantum computing, 6G, and advanced materials, and their effect on the future of Industry 5.0.
- Looking at the expected social and economic changes that will occur as a result of technology in Industry 5.0, such as changes in labor markets and changes in society.
- To give an idea of what the future of Industry 6.0, human–machine integration, artificial general intelligence (AGI), and the role of biotechnology will be.
- To examine the ethical and philosophical questions of advanced technologies and the issues they raise for society.
- To present the concept of scenario planning and foresight as the means for businesses to foresee and be ready for changes in Industry 5.0 and beyond.
- To know about the advantages of new technologies in the development of more effective, environmentally friendly, and human-oriented industrial conditions.

11.2 Emerging Technologies Impacting Industry 5.0

With Industry 5.0, several new technologies will come to the fore as the next wave of innovation. Such technologies are expected to improve processing power, connectivity and material science and allow previously unfathomable levels of optimization, sustainability, and efficiency in manufacturing and beyond. The effect of quantum computing, 6G technology, and advanced materials in the transformation of the industry 5.0 landscape is discussed below.

M. Hammad et al., *Industry 5.0: The Human-Centric Future of Industry*, Springer Series in Advanced Manufacturing, https://doi.org/10.1007/978-3-032-12622-1_11

11.2.1 Quantum Computing

Quantum computing is a paradigm shift towards computational power. In contrast to classical computers, quantum computers operate on quantum bits (qubits) instead of the binary bits used in classical computers to process information. Quits can exist in more than one state at a time, and this allows quantum computers to solve some kinds of problems much faster than can be done by classical systems [1].

- **Impact on Industry 5.0**

  Advanced AI and Machine Learning: Quantum computing will greatly boost AI, and machine learning will be trained much quicker, which will result in better predictive maintenance and optimization of logistics and supply chains.

  Materials Science: The capability of quantum computers to simulate molecular interactions may be used to speed up materials science research, allowing the creation of new materials with novel properties, e.g. stronger and more durable manufacturing components or more efficient energy storage devices.

  Optimization Problems: Quantum computing can be applied to solve real-time complex optimization problems, such as dynamic scheduling, route optimization, and resource allocation and make the manufacturing process efficient and cheaper [2, 3].
- **Examples**: IBM, Google and Microsoft already take into consideration the potential of quantum computing in cryptography, optimization and even artificial intelligence. One application in the pharmaceutical sector is that quantum algorithms could accelerate the process of drug discovery by modeling molecular interactions with a level of detail that cannot be accessed using classical computing.

11.2.2 6G Technology

The upcoming wireless communication technology will be the 6G that is predicted to be offering remarkable data rates, extremely low latency, and global connectivity which is much further than the current 5G networks can offer. 6G will play a crucial role in the functioning of smart cities and self-driving systems [4].

- **Impact on Industry 5.0**

  Improved IoT Network: 6G will offer high bandwidth and low latency which will enable the seamless communication of IoT objects, artificial intelligence and cobots in smart manufacturing plant and will lead to more coordinated and real time transfer of data. This will enhance interaction of human beings and machines and smart automation that can be much more effective and dynamic.

Autonomous Vehicles: 6G will also provide the communication that is needed in real-time to make autonomous vehicles possible, as the vehicles can communicate with each other and the environment to enhance safety and traffic.

Healthcare: 6G will also enable remote surgeries, real-time diagnostics and personalized medicine, which would be enabled by 6G to deliver ultra-reliable, low-latency connectivity to support the complex technologies used in healthcare, which need to interact in real-time [5, 6].

- **Examples**: The 6G can transform traffic in the smart cities by enabling vehicles-to-infrastructure vehicle-to-traffic light communications that would go in real-time and minimize congestion while maximizing the traffic flow. The smart traffic systems might be introduced into the city, and the signals will vary with the traffic that will ultimately lead to an improved scenario in terms of safety and energy consumption.

11.2.3 Advanced Materials

The new materials such as graphene, carbon nanotubes and metamaterials will revolutionize most industries and provide new properties that can significantly enhance strength, durability and energy efficiency. Anything, electronics, automotive parts, etc., can be improved with the help of such materials and the path to more sustainable and innovative products may be opened.

- **Impact on Industry 5.0**

  Energy Efficiency: The technology of batteries can be enhanced with the help of sophisticated materials such as graphene and it will lead to the development of batteries lasting longer and charging quicker. It can accelerate the adoption of electric vehicles (EVs) and renewable energy storage systems and make them less expensive and efficient.

  Lightweight and Strong Parts: Carbon nanotubes and others can be manufactured into light but super strong products. Such industries as the automotive industry and the aerospace industry can use these materials in order to develop stronger and lighter components that are less energy consuming and efficient.

  Sustainable Manufacturing: The other method of sustainable manufacturing can be done using sophisticated materials and the materials will enable generation of products that last longer, and will involve less extraction of raw materials and less waste. These materials also enable the circle economy as they can be recycled and reused [7].
- **Examples**: The other automotive firms such as Tesla are also looking into the use of advanced materials to make light and efficient electric vehicles. The ability to charge faster and longer would allow electric cars to cover a greater distance with the use of graphene-based batteries, cutting carbon emissions and fostering clean energy. Electronics manufacturers are turning to new materials such as graphene

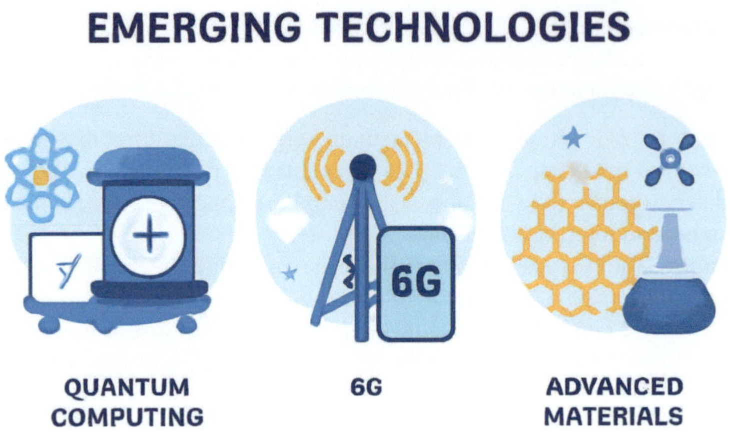

Fig. 11.1 Emerging technologies: driving Industry 5.0's evolution

and metamaterials as their base to build faster and more efficient devices. Such materials allow rapid processors and miniaturization to produce smaller, more powerful, and energy-efficient devices.

The Fig. 11.1 presents a visual representation of three key emerging technologies shaping the future of Industry 5.0: Quantum Computing, 6G Technology, and Advanced Materials. Each technology is depicted through an icon that highlights its core concept: quantum computing with a quantum processor, 6G with a high-speed antenna, and advanced materials with a molecular structure.

Quantum computing, 6G technology, and advanced materials will be major players in changing the industry as Industry 5.0 is introduced. Not only will these new technologies increase processing power, connectivity and material science, but they will also aid the human–machine cooperation that is at the center of Industry 5.0. The future is bright; as these technologies integrate, businesses will achieve optimization, sustainability, and efficiency to an extent never before realized, creating an opening to new and more sustainable industrial activity. To increase AI and machine learning and make new horizons in predictive maintenance, drug discovery, and climate modelling is the future of quantum computing that will ensure that there will be no bottlenecks to obstruct the creation of smart cities, autonomous systems, and advanced healthcare. Advanced materials will also employ innovative technologies to create energy efficient, lightweight structures and sustainable manufacture to create more environmental friendly products and solutions. The future of industry 5.0 will be, when these technologies will be balanced with human abilities of creativity, moral responsibility, and sustainability and offer efficient, innovative, and ecologically-friendly industrial environment.

11.3 Social and Economic Impacts: Potential Shifts in Labor Markets and Societal Structures

Industry 5.0 will lead to drastic shifts in the social and economic order with the implementation of new high-tech technologies, automation, and human-centered design. The appearance of the AI, robotics, IoT, and the innovations that are oriented to sustainability will redefine the structure of the work, the skills required as well, and the overall framework of the society. The changes in the labor markets and social norms will be caused by the technological advancement and the character of work.

11.3.1 Changes in the Labor Markets

The incorporation of high technologies into industry will result in considerable changes in the labor markets. Some of the old jobs will be substituted with others that are innovative, highly skilled and creative.

- **Displacement and creation of Jobs**: Although Industry 5.0 will introduce automation of routine work by means of robotization and AI, it will also lead to the emergence of a new type of jobs aimed at more specific and high-skill employment. Such new jobs will be in areas such as AI management, robotics programming, data science, cybersecurity, and advanced manufacturing systems.

  Key Insights

  - The displacement of jobs would occur in jobs that are routine, manual, and monotonous.
  - The creation of jobs will be focused on such domains as the development of AI, machine learning, robotics engineering, data analysis, and sustainability-related employment.
  - Creativity and adaptability will become the major assets in the workforce because a human will be working closer with a machine in more strategic functions.

  An example is that in the manufacture of automobiles, robots have replaced such activities as welding and painting. New occupations have appeared in the field of robotics programming, AI maintenance and quality control, and customer service jobs that cannot be automated without human imagination to solve problems.
- **Skill Gaps**: The high rate of technological change is bound to cause a skills gap, in which the need for higher technological skills will be more than the current workforce can deliver. To solve this, organizations will be required to invest in ongoing upskilling and reskilling schemes for their employees.

  Key Insights

  - The current labor force will have to be reskilled in order to operate with the latest technologies, including AI and robotics.
  - Businesses will have to work together with the education systems to make sure that the future workforce acquires the skills they need.
  - Continuous professional development and lifelong learning will be an essential part of career development.

  Example: Volvo provides training to its employees in digital tools, AI and robotics. This will guarantee that the workers may cooperate with independent systems on the production line instead of being substituted by them.

- **More Remote Work**: With the change in the emphasis of Industry 5.0 to automation of routine duties, the employees in the knowledge-based economy, including technology, consulting, and design, will enjoy greater freedom to work at a distance. This will give the workers greater freedom and the opportunity to indulge in more creative and strategic activities.

  Key Insights: Automation of routine activities will save time on manual labor, and workers will focus on high-value activities.

  - It will increase the demand of working remotely, particularly in sectors where employees are not required to be physically present in order to work.
  - The better digital communication technology and cloud computing will help to move to a more work-at-home force [8].

  Example: Remote-first consulting firms and technology companies are also enabling their employees to work at home or anywhere provided all the tedious administrative tasks are still done by automation.

11.3.2 Societal Changes

Industry 5.0 will also be a cause of significant transformation to the social fabric. As more technology is being incorporated in daily living, equity, work-life balance, and universal access to opportunities will become the most significant societal changes.

- **Equity and Inclusion**

One of the potential benefits of Industry 5.0 is the possibility of bridging the social gap that the current supply of new technologies pits people against each other and making the workplace more accommodating to everyone involved. However, the other risk is that the marginalized people would be left behind in case they are unable to get the required training, education or technological solutions to make it in this new economy.

Key Insights

- Access to technology and education will be paramount to avoid the further development of inequalities.

- The use of Industry 5.0 technologies must be done in a manner that will facilitate the establishment of inclusive workplaces and enable marginalized groups to enjoy the benefits of the expanding digital economy.
- Governments, businesses and education systems will have to collaborate to give opportunities to everyone.

Example: Estonia has emerged as a digital government and inclusive access to technology leader. The e-resident program of the country has enabled people all over the world to use digital services, and this has opened up economic opportunities to people in areas where they cannot access traditional markets.

- **Universal Basic Income (UBI)**

Industry 5.0 is characterized by the increasing automation and use of robotics, and it can be argued that this can cause the displacement of workers, so the question arises whether Universal Basic Income (UBI) could become one of the viable solutions to help those who lost their jobs. In UBI, every citizen gets a certain amount of money as an income that is independent of the job.

Key Insights

- Since automation is cutting down the need for human labor in some areas of work, UBI might provide people with a minimum amount of income to be able to sustain themselves and their families.
- UBI may also be used to deal with wealth inequality by serving as a safety net for people who lose their jobs to technology.
- Case study: A pilot of UBI in Finland had positive outcomes, as it gave its participants financial stability and freedom to either continue with their education or start their business without the need to have a regular job.

- **Work-Life Balance**

The focus on human–machine interaction and automation of routine activities in the context of Industry 5.0 will allow workers to have an improved work-life balance. Since more routine jobs are performed by automation, the workers can do more strategic, creative and higher-level work, which is both more satisfying to the individual and intellectually challenging [9].

Key Insights: Automation will help employees to spend more time on more meaningful tasks, and they will have better well-being.

- The accent on individuality and adaptability will enable employees to match their careers to their personal ideals and life preferences.
- Increased freedom and the potential to work at their convenience will enhance the satisfaction rate of the employees and reduce the amount of stress that comes with a conventional office set up.
- Case study: Salesforce is one of the companies that are concerned with the work-life balance and well-being of the employees. Replacement of repetitive tasks

by using AI-powered systems will assist employees to focus on innovation and self-development that will boost productivity and job satisfaction.

Figure 11.2 depicts numerous social and economic changes that are expected during the times of Industry 5.0. It shows how workers can deal with the robots, which is a manifestation of how automation and AI affect the labor market. Job displacement, and the establishment of the new lab opportunities in AI, data science, and robotics are mentioned. The picture also depicts the notions of skills shortages, remote working more often, equity and inclusion, and the possibility of universal basic income as remedies to displaced workers. Moreover, routine work is implied to be automated to gain better work-life balance.

Fig. 11.2 Automation and equity: shaping the workforce of tomorrow

11.3.3 Advantages

- **Economic Growth**: Industry 5.0 has the potential to spur the overall economic growth because it will encourage new industries, innovation, and new high-tech manufacturing systems. The growth will be based on the areas of AI, sustainability, and data science.
- **Workers Empowerment**: With the increasing level of innovation-based work and human–machines collaboration, employees will no longer be afraid of pursuing intellectually engaging employment opportunities. This may lead to job satisfaction, engage and independence, which have immense importance in the long-term economic welfare.
- **Sustainability**: Industry 5.0 will contribute to sustainable economic development, and it will be correlated to the technological advancement and green technologies and social justice. This will assist the businesses and the society to have more sustainable practices that will have minimal impact on the environment.

With the introduction of Industry 5.0, which brings a new age of human-centered technologies, social and economic effects will be enormous. As job displacement and skill shortage are among the major issues, continuous training and reskilling of the workforce will become essential. Meanwhile, new employment opportunities, increased creativity, and work-life balance will give workers more freedom and satisfaction. UBI and equity programs will gain more significance in the context of the possible social inequalities that can emerge. Industry 5.0 has the possibility of building a more balanced, equitable and resilient economy in the future by adopting the concept of sustainability and inclusivity [10].

11.4 Future Visions: Industry 6.0 and Beyond

Since Industry 5.0 is the next step in the development of human–machine cooperation, where the emphasis is put on creativity, sustainability, and ethical responsibility, the path of technological progress is already pointing at the development of Industry 6.0. The shift to Industry 6.0 is catalyzed by the innovations of quantum computing, neural interface, advanced robotics, synthetic biology, and artificial general intelligence (AGI) that may change the relationship between people and technology in a way never seen before.

11.4.1 Industry 6.0: The Human–Machine Integration

Industry 6.0 may be the next step in human–machine synergy and go beyond cooperation to integration. With more advanced machines, human capacities may be expanded by using direct brain–computer interfaces (BCIs) to allow more natural and direct interaction between humans and machines.

- **Complete Brain–Computer Interfaces (BCI)**: The industry 6.0 BCI might allow human brains to communicate directly with machines. This would essentially transform the manner in which humans carry out their duties as they would be in a position to operate machines, robots, and even the whole production system using their thoughts only.
- **Industry Effect**: Employees would be able to operate complicated machines, robots, and AI without physical interfaces such as keyboards, joysticks, or voice recognition. This may result in the radical transformation of the workplace, where human employees may be able to concentrate solely on creative problem-solving and strategic decision-making, whereas AI and robotic systems perform the physical work. The pace and effectiveness of human–machine interactions may also be enhanced by several folds, providing greater levels of productivity or even allowing the workers to multitask or work on several machines at the same time.

  Example: A factory worker might be able to operate a robot doing work in the production line with just a thought in his head and concentrate his/her mind on more important decisions or innovations.
- **Artificial General Intelligence (AGI)**: Whereas existing AI systems are limited in terms of what they can do, AGI is the next step in developing AI systems that have generalized human-like intelligence. This would enable machines to comprehend, learn and do a variety of tasks and adapt to new challenges without having to be trained on each one of them.
- **Industry Effect**: AGI systems may be able to operate independently in problem-solving, real-time innovation, and multitasking in different industries. AGI could strengthen human–machine cooperation by independently performing those tasks that could not be predicted or were too complicated to be done by the existing AI systems. The AGI would probably result in the streamlining of business processes, be it manufacturing, research or even creative businesses such as design or entertainment.

  Example: AGI systems may be able to autonomously control a whole supply chain, predicting disruptions, changing the logistics and optimizing production processes in real-time.

11.4.2 The Place of Biotechnologies and Synthetic Biology

Amid the increasing calls to be more sustainable, biotechnology and synthetic biology will become a significant part of Industry 6.0 and will alter the way industries obtain materials and manufacture products. Such biotechnological innovations may transform the manufacturing process to be environmentally friendly and biologically incorporated.

- **Advanced Biotechnology**: Biotechnology will allow industries to use bio-based materials and bio-manufacturing processes, which will allow minimizing the impact on the environment and developing a more sustainable industrial ecosystem.
- **Industry Effect**: Traditional manufacturing of products being replaced by biological manufacturing of products might be a substitute of chemical or mechanical manufacturing by bio-manufacturing. Biodegradable materials, such as biopolymers, can be incorporated in the making of the products to make it eco-friendly products that do not require high reliance on non-renewable materials. Bio-based production may create a vast number of products, such as medicines and consumer goods, using renewable biological resources rather than synthetic chemicals.

  Example: Algae or plants-derived bio plastics could replace petroleum-based ones, which would reduce environmental pollution.
- **Synthetic Biology**: The next stage in genetic engineering is synthetic biology, in which custom-designed microorganisms are created that can be used to create chemicals, fuels and materials that are synthesized more effectively than through the normal manufacturing process.
- **Industry Effect**: Biosynthesis of materials can change the manufacturing process, with microorganisms being built to produce significant raw materials (or even finished products) in an energy-efficient method. The latter can lead to more eco-friendly production processes that use less energy and fewer chemicals and generate minimum wastes.

  Example: Other firms, including Ginkgo Bioworks, already create custom microbes capable of making everything, including ingredients and pharmaceuticals, a potentially highly sustainable and efficient alternative to conventional chemical manufacturing.

11.4.3 Ethical and Philosophical Implications

By introducing even more advanced AI and biotechnologies, Industry 6.0 will become a genuine ethical, philosophical issue of machine place in the society. There will be the necessity to resolve such crucial matters as privacy, autonomy, and the nature of work.

- **Autonomy and AI Rights**: As machines grow more autonomous and as machines may attain cognitive capabilities, the rights and liabilities of the AI systems will become an issue.
- **Fair Allocation of Technology**

Technology will keep increasing and the society must ensure that the gains are shared with them. There is also a possibility that the technological developments will heighten disparities in the event that only a few individuals are given the privilege to savor the fruits of advanced technologies [11].

Fig. 11.3 BCI technology: the next step in human–machine synergy

Example: The governments may be forced to implement legislations to ensure that all people have access to AI and brain–computer technologies regardless of their economic status and their location.

Figure 11.3 represents a hypothetical picture of Industry 6.0 in which human–machine integration will be smooth. It depicts a human and a robot in a laboratory setting, which puts the emphasis on the high-tech brain–computer interface (BCI) technology that enables direct communication between the mind of a human and the machine. The background shows another scientist who is working with high technological equipment. The scene represents the possible future of the industries with the collaboration of human cognition, robotics, and artificial intelligence in new unimaginable ways.

11.4.4 Advantages of Industry 6.0

- **Increased Efficiency**: AI, robotics and biotechnologies that will emerge in Industry 6.0 shall lead to unprecedented levels of efficiency and productivity that will overhaul entire industries. The ease of work processes and the ability of AI to independently solve problems will make the industrial environment a smooth and highly efficient one.
- **Improved Human–Machine Synergy**: Human–machine integration will make it possible to have a more dynamic cooperation and it will be capable of incorporating human creativity and machine efficiency. This kind of synergy can lead to breakthrough in production of medicine industries.

- **Revolutionary Innovations**: Industry 6.0 can bring new innovations to such fields as healthcare, sustainability, and energy production. The opportunities of new more sustainable materials and efficient energy sources are almost limitless when such technologies as artificial biology and quantum computerization are developed and contend to help solve some of the most burning problems in the world [12].

11.5 Scenario Planning and Foresight Methodology

As Industry 5.0 and future evolves, businesses and governments must adopt scenario planning and foresight strategies that would assist them to surmount the uncertainties and opportunities of new technology. The tools are useful in providing an idea of what is likely to occur in the future hence the organizations can be proactive in their strategies and prepared to meet various eventualities. This can be done by the use of these analytical frameworks enabling the organizations to be in a position to succeed in a more dynamic environment.

11.5.1 Scenario Planning

Scenario planning is a planning method that assists one to explore different ways things may unfold based on key change agents. It helps organizations envision different ways that the different trends, uncertainties and emerging technologies can alter their industries hence enabling them to develop adaptive plans that can respond to the evolving circumstances.

Scenario planning usually entails the identification of critical uncertainties—factors that may develop in various ways and have a great impact on the future of an organization. This may be in the form of AI, quantum computing or sustainability regulations. Organizations then develop several reasonable scenarios that investigate how these uncertainties would unfold with time. It is about the development of a variety of potential futures, not trying to foresee a certain outcome. The aim is to create strategic flexibility and make the organization ready to seize unplanned disturbances or opportunities as they come.

- **Example**: Scenario planning can help a manufacturing company assess the possible impact of automation through AI and quantum computing on its activities in the next 10–20 years. The company could investigate:

  Scenario 1: The future where AI and quantum computing have been widely used and will result in more automation, lower cost of production, and shift towards very personalized products.

Scenario 2: A situation where these technologies will be embraced gradually due to the regulatory restrictions and the company will be forced to remain with the traditional manufacturing methods with slight improvement on efficiency.

Investigating different situations, the company will be capable of devising certain flexible approaches, such as investing in training its employees on AI technologies or introducing new sets of products that would be less automated, depending on which scenario is more probable to occur.

- **Advantages**: The advantages of Scenario Planning:

  Strategic Agility: Scenario planning can also render business agile and flexible so that they adapt to changes when situations change.

  Informed Decision-Making: It allows organizations to decide in a long-term perspective and consider other potential future possibilities rather than addressing prevailing issues.

  Futureproofing: Scenario planning may help companies in identifying potential threats and opportunities long before their time, therefore, they become better equipped to address disruptive changes [13].

11.5.2 Foresight Methodologies

Foresight methodologies can be seen as a methodical means of finding, examining and valuing the new trends and technologies. These strategies enable the organizations to determine the signs of what is expected in the future and ensure that they stay ahead of the game and align their strategies accordingly.

The most important Foresight Methodologies:

- **Trend Analysis**: Trend analysis involves the examination of the present trends in the field of technology, society, economy, and politics in an attempt to forecast the future. By identifying the key trends, companies will be given the opportunity to project how they can evolve and transform their sector.

  Example: A car company might analyze the trend in the increase of electric vehicles (EVs), self-driving cars, and sustainability law to predict the future of the car industry in terms of how this shift in trends will transform the automotive world.

- **Professional Seminars and Discussions**: Expert workshops are the gatherings of industry leaders, technologists, and researchers who discuss future challenges and opportunities. These workshops enable organizations to tap into the pool of knowledge of thought leaders to be more equipped with emerging technologies and possible disruptions.

  Example: An expert workshop on the topic of how AI-driven governance, 5G networks, and smart infrastructure will change the way people live in cities may take place in the framework of a smart city project. It may be possible to use the insights acquired to make strategic decisions about the development of smart cities.

- **Technology Road mapping**: Technology roadmaps assist organizations in the process of mapping the development of some of the major technologies and determining how they can affect business operations. Road mapping makes businesses see the future of the technologies and create plans accordingly [14].

  Example: An example is that a medical practitioner can use technology road mapping to monitor the change in telemedicine, wearable health gadgets, and AI-based diagnostics. Mapping such technologies will enable the provider to have a strategic plan on how to incorporate them into future services.

11.5.2.1 Foresight Methodology Example

Foresight methodologies might be applied to a smart city project to determine the future of 5G, AI-based governance, and smart infrastructure in the creation of cities. The project team might also be able to find the most promising innovations and decide which technologies are to be prioritized in order to be integrated through expert workshops and technology road mapping. Through foresight, the city will be ready to face future issues that may include traffic congestion, energy management, and urban mobility.

11.5.2.2 Foresight Methodologies: Advantages

- **Early Detection of Opportunities and Threats**: Foresight practices also help the firm to identify the emerging threats and opportunities before they mainstream and therefore organizations can adapt and innovate.
- **Informed Strategic Planning**: The tools are significant in providing a solid foundation upon which strategic decision-making process can be based in such a way that organizations can plan their resources with the new technological changes and trends.
- **Competitive Advantage**: Organizations which apply foresight in order to predict trends and technological shocks become competitive as they are the first ones to take advantage of a new opportunity or to prevent possible risks.

11.6 The Combination of Scenario Planning and Foresight

Scenario planning and foresight approaches are considered a holistic way of future-proofing organizations. The two concepts of scenario planning (the capacity to visualize various futures) and foresight (the insights provided by emerging trends), when

Fig. 11.4 Anticipating the future with scenario planning and foresight

combined, allow businesses to create solid strategies that can be modified to meet a broad variety of future possibilities.

Example: An AI and robotics technology company may have to integrate scenario planning and foresight techniques to develop a decade-long strategy. Scenario planning may help the company explore the potential futures with regard to the regulatory landscape of AI and global competition, but foresight methods would help the company explore the future changes of quantum computing and biotech integration.

A combination of these approaches would allow the firm to develop an active product development, research capital, and market growth plan, becoming successful in the dynamic technological world.

The ideas of scenario planning and foresight methodologies are shown in Fig. 11.4. Scenario planning will have an icon of a man thinking to represent imagining alternative futures that can happen. The methodologies of foresight are symbolized by an icon of a telescope, which symbolizes the application of instruments to predict the imminent trends and technologies. The picture shows the significance of these techniques in future planning of industries especially in Industry 5.0 and more.

11.6.1 Advantages

The benefits of using scenario planning and foresight methods:

- **Proactive Decision-Making**: The tools allow organizations to be proactive and not reactive. With the consideration of the range of possible futures available, businesses will be capable of making good decisions that will assist them to overcome the uncertain and seize the emerging opportunities.

- **Improved Risk Management**: Scenario planning and foresight helps the business to identify the risks they are exposed to and as such come up with measures to mitigate risks before they develop into a huge problem.
- **Strategic Flexibility**: The strategies advocate strategic flexibility, which assists an organization to change or adapt to the new trends, technologies, or even challenges.
- **Informed Long-Term Strategy**: Scenario planning and foresight also have a systematic approach to planning in the long term in which the organizations are prepared to respond to disruptions and also able to align their resources and efforts to the way the industry will develop.

The scenario planning and foresight technique is extremely essential in the development of organizations within the constantly evolving environment of the transitioning industry 5.0 and future shocks. These approaches enable companies to explore every potential outcome and identify the new trends which is a definite approach to proactive decision-making, risk-taking, and flexibility of strategies. The tools are useful in a world where fast technological change and global uncertainty are the new norm, and organizations have to be at the forefront of the change to succeed [15].

11.7 Learning Outcomes

At the end of this chapter, the reader will be capable of:

- Learn about the new technologies that are likely to define Industry 5.0 and Industry 6.0, such as quantum computing, 6G and advanced materials.
- Determine the social and economic effects of these technologies that may be experienced in the future, particularly with regard to labor markets and the structures of society.
- Discover the ethical issues and opportunities of human–machine integration and AI in future industries.
- Use scenario planning and foresight techniques to plan and get ready for what should happen in the future of Industry 5.0.
- Explain how these new technologies would affect sustainability, innovation and human-friendly practices in different industries.
- Learn about the position of industries in spurring innovation by means of technological development with related issues and obstacles.

11.8 Review Questions

1. (a) What role will quantum computing play in strengthening the powers of AI and machine learning in Industry 5.0?

   (b) Explain how it could be used in predictive maintenance, logistics and climate modelling and what makes it hard to become widely accepted.

2. (a) What will 6G do to increase the speed of interconnectivity of IoT devices, cobots, and AI systems in smart manufacturing?

   (b) Give examples of how it can affect real-time data exchange and human–machine collaboration.
3. (a) What will be the impact of graphene and carbon nanotubes on the manufacturing process, efficiency of products and sustainability?

   (b) Briefly describe their uses in industries such as automotive, electronics and energy.
4. (a) What are the prospects for job displacement and job creation in this new industrial era?

   (b) Write about the competencies that the workforce of the future will need and how businesses can gear up workers to meet the emerging job profiles?
5. (a) What is the contribution of Industry 5.0 to equity and inclusion in society?

   (b) What can organizations do to make sure that disadvantaged groups can enjoy the benefits of technological breakthroughs, and what are the threats of further inequality in society?
6. (a) What are the possible advantages and disadvantages of the Universal Basic Income (UBI) as the solution to job displacement in Industry 5.0?

   (b) Explain the possibility of UBI and how it will affect the future labor market and economic systems.
7. (a) What are the ways through which the change in labor dynamics can bring in more empowerment and creativity of workers?

   (b) Give examples of how automation can introduce more fulfilling jobs in the workplace.
8. (a) What will be the next step of human–machine collaboration and decision-making due to BCI and AGI?

   (b) Which ethical issues are going to take place with the advancement of these technologies?
9. (a) What are the ways that industries can employ scenario planning and foresight techniques to anticipate the shocks and the uncertainties of the future?

   (b) Explain how companies can use these tools to prepare themselves for the possible effects of quantum computing, 6G, and other upcoming technologies.
10. Sustainable production may be transformed by incorporating biotechnologies and synthetic biology in the production process.

    (a) What are the ways in which these developments can help to minimize the environmental impact of the industries?

    (b) What are the difficulties which businesses will have encounter in order to implement these technologies?

Appendix: Solution Manual

1. **(a) What role will quantum computing play in strengthening the powers of AI and machine learning in Industry 5.0?**

Solution: Quantum computing can enable AI and machine learning to process vast amounts of data much faster, leading to more accurate predictive models.

(b) Explain how it could be used in predictive maintenance, logistics and climate modelling and what makes it hard to become widely accepted.

Solution: In predictive maintenance, it can identify failure points in machinery faster than classical computing, reducing downtime. However, quantum computing's widespread adoption faces challenges in terms of hardware limitations and the need for highly specialized skills to develop quantum algorithms.

2. **(a) What will 6G do to increase the speed of interconnectivity of IoT devices, cobots, and AI systems in smart manufacturing?**

Solution: 6G technology will enhance the interconnectivity of IoT devices, cobots, and AI by providing ultra-low latency and high-speed data transmission, allowing real-time collaboration in smart manufacturing.

(b) Give examples of how it can affect real-time data exchange and human–machine collaboration.

Solution: For example, autonomous vehicles on a production line can communicate with AI systems to adjust production speeds or redirect materials, significantly improving efficiency. Challenges include the infrastructure needed to support 6G and ensuring global standardization.

3. **(a) What will be the impact of graphene and carbon nanotubes on the manufacturing process, efficiency of products and sustainability?**

Solution: Advanced materials like graphene and carbon nanotubes offer superior strength and conductivity, leading to more energy-efficient products.

(b) Briefly describe their uses in industries such as automotive, electronics and energy.

Solution: In automotive manufacturing, these materials can reduce vehicle weight and improve fuel efficiency. In electronics, they allow for faster, smaller, and more efficient devices. Challenges include scaling production and ensuring cost-effectiveness.

4. **(a) What are the prospects for job displacement and job creation in this new industrial era?**

Solution: As automation and AI reduce the need for manual labor, new roles requiring technical, creative, and strategic skills will emerge. Workers will need skills in AI management, robotics programming, and data analysis.

(b) Write about the competencies that the workforce of the future will need and how businesses can gear up workers to meet the emerging job profiles?

Solution: Workers will need skills in AI management, robotics programming, and data analysis. Companies can prepare by investing in upskilling programs and fostering adaptability. This transition can lead to a more intellectually engaging workforce, although it may also lead to job displacement for lower-skilled workers.

5. **What is the contribution of Industry 5.0 to equity and inclusion in society?**

Solution: Industry 5.0 could promote equity and inclusion by providing equal access to technological tools and creating jobs that require creative problem-solving, which can empower marginalized groups.

(b) What can organizations do to make sure that disadvantaged groups can enjoy the benefits of technological breakthroughs, and what are the threats of further inequality in society?

Solution: If access to education and technology is not equally distributed, it could exacerbate social inequality. Companies must ensure inclusive policies and training programs to prevent this.

6. **(a) What are the possible advantages and disadvantages of the Universal Basic Income (UBI) as the solution to job displacement in Industry 5.0?**

Solution: UBI could provide a safety net for workers displaced by automation, ensuring economic stability. However, the feasibility of UBI depends on the sustainability of the economy and the capacity of industries to generate wealth in a highly automated future.

(b) Explain the possibility of UBI and how it will affect the future labor market and economic systems.

Solution: Critics argue that UBI might discourage work and lead to inflation, while proponents suggest it could reduce poverty and income inequality.

7. **(a) What are the ways through which the change in labor dynamics can bring in more empowerment and creativity of workers?**

Solution: By automating repetitive tasks, workers can focus on higher-level creative roles, leading to more fulfilling work. For example, AI could handle data analysis, allowing human workers to engage in strategy and innovation.

(b) Give examples of how automation can introduce more fulfilling jobs in the workplace.

Solution: For example, AI could handle data analysis, allowing human workers to engage in strategy and innovation. The challenge is ensuring that workers have the skills to take on these new roles, which requires investment in training and education.

8. **(a) What will be the next step of human–machine collaboration and decision-making due to BCI and AGI?**

Solution: BCIs and AGI will enable more intuitive and efficient human–machine collaboration, where humans can control machines directly with their thoughts, significantly enhancing productivity and creativity.

(b) Which ethical issues are going to take place with the advancement of these technologies?

Solution: Ethical concerns include privacy issues related to brain data and questions around AI's autonomy and decision-making rights. Regulations and ethical guidelines will be necessary to ensure that these technologies are used responsibly.

9. **(a) What are the ways that industries can employ scenario planning and foresight techniques to anticipate the shocks and the uncertainties of the future?**

Solution: Scenario planning allows businesses to explore multiple future outcomes based on technological advancements and market changes. For example, a company might explore how quantum computing could transform their supply chain or how 6G will impact real-time customer service.

(b) Explain how companies can use these tools to prepare themselves for the possible effects of quantum computing, 6G, and other upcoming technologies.

Solution: For example, a company might explore how quantum computing could transform their supply chain or how 6G will impact real-time customer service. These methodologies help businesses adapt to rapid technological changes by providing a framework for flexible decision-making.

10. **Sustainable production may be transformed by incorporating biotechnologies and synthetic biology in the production process.**

(a) What are the ways in which these developments can help to minimize the environmental impact of the industries?

Solution: Biotechnologies and synthetic biology can lead to sustainable manufacturing by enabling the production of bio-based materials, reducing reliance on non-renewable resources. These technologies can help create more energy-efficient products and reduce waste.

(b) What are the difficulties which businesses will have encounter in order to implement these technologies?

Solution: Challenges include the need for substantial research and development and the potential ethical concerns related to genetic manipulation.

References

1. Khan WA et al (2025) Quantum computing—computer aided engineering design and manufacturing: a fourth industrial revolution perspective. Springer, Cham, pp 537–549

2. Lu Y et al (2023) Quantum computing and industrial information integration: a review. J Ind Inform Integr 35:100511
3. Balarabe K (2025) Quantum computing and the law: navigating the legal implications of a quantum leap. Eur J Risk Regulat 15:1–20
4. Haque S, Singh SK (2025) Impact of 6G technologies on Industry 5.0. In: Proceedings of the 6G impacts on natural habitats and human life. IGI Global Scientific Publishing, New York, pp 109–134
5. Rojek I et al (2024) Review of the 6G-based supply chain management within industry 4.0/5.0 paradigm. Electronics 13(13):2624
6. Noor-A-Rahim M et al (2022) Toward industry 5.0: intelligent reflecting surface in smart manufacturing. IEEE Commun Mag 60(10):72–78
7. Vashi D, Sultan M (2025) AI in quantum computing: exploring synergies and possibilities
8. Adel A, Alani NHS, Jan T (2024) Factories of the future in industry 5.0—softwarization, servitization, and industrialization. Internet Things 28:101431
9. Hammad M et al (2025) Future trends and innovations in smart manufacturing. Smart manufacturing blueprint: navigating industry 4.0 across diverse sectors. Springer, Cham, pp 413–438
10. Rejeb A et al (2025) When Industry 5.0 meets the circular economy: a systematic literature review. Circ Econ Sustain 16:1–32
11. Khan WA et al (2025) Nexus of CIM, Industry 4.0, Industry 5.0, and quantum computing. Computer aided engineering design and manufacturing: a fourth industrial revolution perspective. Springer, Cham, pp 551–557
12. Hammad M et al (2023) Framework for the implementation of smart manufacturing systems: a case in point. Processes 11(5):1436
13. Hammad M et al (2025) Crafting a sustainable and innovative future. Smart manufacturing blueprint: navigating Industry 4.0 across diverse sectors. Springer, Cham, pp 441–469
14. Awotunde JB et al (2025) Advancements in computational intelligence for enhanced big data analytics in Industry 4.0 and 5.0: challenges, advances, and future perspectives. Computational intelligence for analysis of trends in Industry 4.0 and 5.0. Auerbach Publications, pp 39–84
15. Barata J, Kayser I (2023) Industry 5.0—past, present, and near future. Proced Comput Sci 219:778–788

MIX
Papier aus verantwortungsvollen Quellen
Paper from responsible sources
FSC® C105338

If you have any concerns about our products,
you can contact us on
ProductSafety@springernature.com

In case Publisher is established outside the EU,
the EU authorized representative is:
Springer Nature Customer Service Center GmbH
Europaplatz 3, 69115 Heidelberg, Germany

Printed by Libri Plureos GmbH
in Hamburg, Germany